ENCYCLOPEDIA OF
BODY

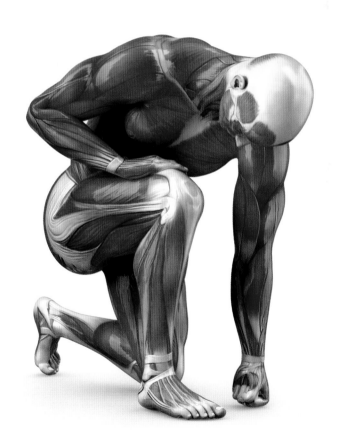

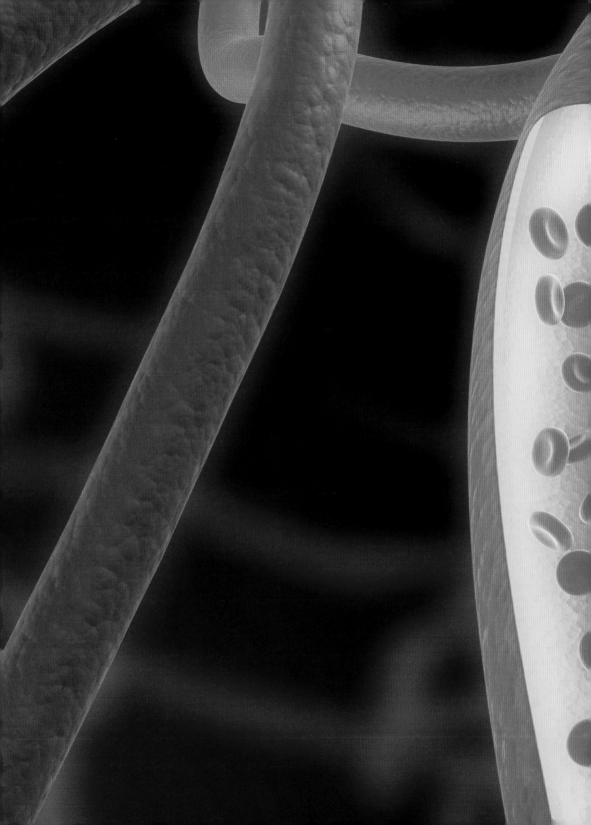

ENCYCLOPEDIA OF
BODY

John Farndon
Nicki Lampon (Tiger Media)

Miles
Kelly

First published in 2010 by Miles Kelly Publishing Ltd
Harding's Barn, Bardfield End Green, Thaxted, Essex, CM6 3PX, UK

This edition printed in 2018

2 4 6 8 10 9 7 5 3 1

Publishing Director Belinda Gallagher
Creative Director Jo Cowan
Managing Editor Rosie Neave
Assistant Editor Claire Philip
Cover Designer Simon Lee
Series Designer Helen Bracey
Volume Designer Martin Lampon (Tiger Media)
Junior Designer Kayleigh Allen
Image Manager Liberty Newton
Indexer Indexing Specialists (UK) Ltd
Production Elizabeth Collins, Caroline Kelly
Reprographics Stephan Davis, Jennifer Cozens
Assets Lorraine King

ISBN 978-1-78617-432-1

Printed in China

British Library Cataloguing-in-Publication Data
A catalogue record for this book is available from the British Library

Made with paper from a sustainable forest

www.mileskelly.net

Contents

Body structure

Musculoskeletal system

Nervous system

Circulatory system

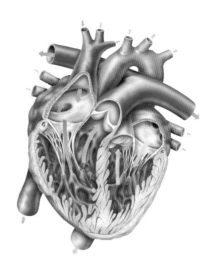

 # Immune system

 # Respiratory system

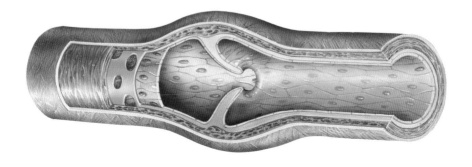

Digestive and urinary systems

Hormones and metabolism

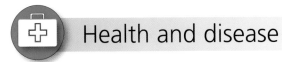

Health and disease

Body structure

The human body

- **There are nearly seven billion people** in the world and we are all different.

- **The human body** is a fascinating and complicated structure made up of millions of cells that all work together.

- **These cells** combine to make tissues. Two or more tissues combine to make organs and several organs combine to make body systems.

- **These systems** are all controlled by complicated interactions of signals and chemicals passing around the body.

Hair colour can vary dramatically

Some people need glasses to see clearly

Just over 50 percent of the population is female

- **Humans are different** from all other animals and have adapted to live all around the world.

- **We are the only** animals to walk upright, allowing us to use our hands for many things.

- **We have** very developed brains and can work out problems, communicate in many ways and enjoy cultural activities, such as music and art.

- **We love and hate**, laugh and cry. We are capable of many emotions and are constantly exploring the world around us.

- **Our bodies** come in all shapes and sizes. One of the world's tallest men is 2.36 m tall, while one contender for the title of world's shortest man is currently only 50.8 cm tall.

We all have an individual skin colour

We can be tall or short, fat or thin

◀ *Human beings are all individual. Apart from identical twins, no two people are alike.*

Body systems

Your body systems are interlinked – each has its own task, but they are all dependent on one another.

The skeleton supports the body, protects major organs, and provides an anchor for the muscles.

The nervous system is the brain and the nerves – the body's control and communications network.

The digestive system breaks down food into chemicals that the body can use to its advantage.

The immune system is the body's defence against germs. It includes white blood cells, antibodies and the lymphatic system.

Water balance inside the body is controlled by the urinary system. This removes extra water as urine and gets rid of impurities in the blood.

The respiratory system takes air into the lungs to supply oxygen, and lets out waste carbon dioxide.

The reproductive system is the smallest of all the systems. It is basically the sexual organs that enable people to have children. It is the only system that is different in men and women.

Other body systems include the hormonal system (controls growth and internal co-ordination by chemical hormones), integumentary system (skin, hair and nails), and the sensory

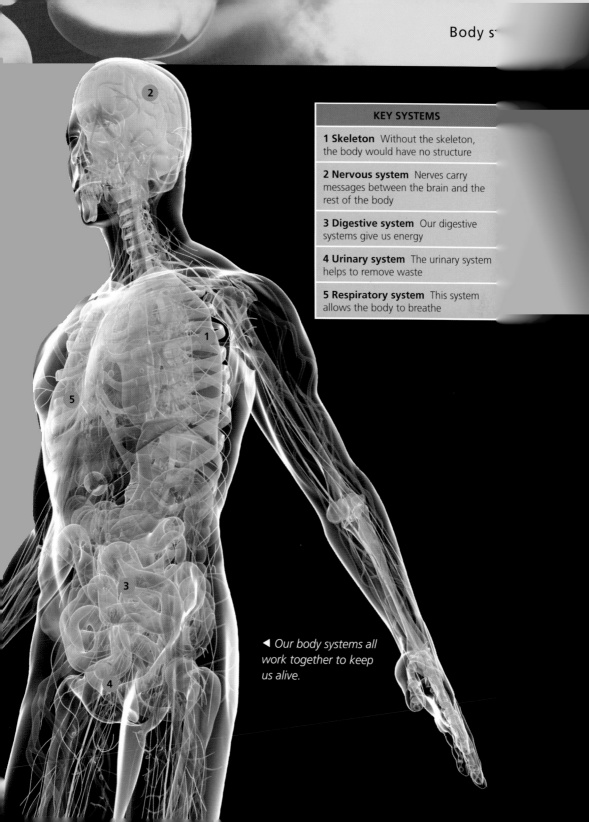

KEY SYSTEMS

1 Skeleton Without the skeleton, the body would have no structure

2 Nervous system Nerves carry messages between the brain and the rest of the body

3 Digestive system Our digestive systems give us energy

4 Urinary system The urinary system helps to remove waste

5 Respiratory system This system allows the body to breathe

◀ *Our body systems all work together to keep us alive.*

Anatomy

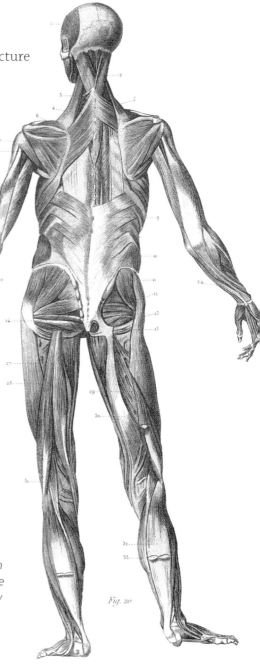

- **Anatomy is the study** of the structure of the human body.

- **Comparative anatomy** compares the structure of our bodies to those of animals' bodies.

- **The first great anatomist** was the ancient Roman physician, Galen (AD 129–199).

- **The first great book** of anatomy was written in 1543 by the Flemish scientist Andreas Vesalius (1514–1564). It is called *De Humani Corporis Fabrica* ('On the Fabric of the Human Body.')

- **In order to describe** the location of body parts, anatomists divide the body into quarters.

▶ *Much of our basic knowledge of human anatomy comes from the anatomists of the 16th and 17th centuries, who meticulously cut up corpses and then accurately drew what they saw.*

Fig. 20

- **The anatomical position** is the way the body is positioned to describe anatomical terms – upright, with the arms hanging down by the sides, and the eyes, palms and toes facing forwards.

- **The central coronal plane** divides the body into front and back halves. Coronal planes are any slice across the body from side to side, parallel to the central coronal plane.

- **The ventral** or anterior is the front half of the body.

- **The dorsal** or posterior is the back half of the body.

- **Every part** of the body has a Latin name, but anatomists use a simple English name if there is one.

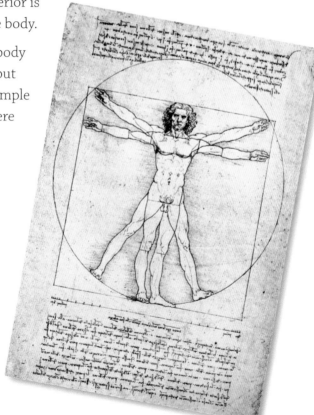

▶ *This drawing by Leonardo da Vinci is called* Vitruvian Man *and shows a man of perfect proportions.*

Tissues

- **A tissue** is a body substance made from many of the same type of cell. Muscle cells make muscle tissue, nerve cells form nerve tissue, and so on.

- **As well as cells**, some tissues include other materials.

- **Connective tissues** are made from particular cells (such as fibroblasts), plus two other materials – long fibres of protein (such as collagen) and a matrix. Matrix is a material in which the cells and fibres are set like the currants in a bun.

- **Connective tissue** holds all the other kinds of tissue together in various ways. The adipose tissue that makes fat, tendons and cartilage is connective tissue.

- **Bone and blood** are both connective tissues.

▼ Lungs are largely made from special lung tissues, but the mucous membrane that lines the airways is epithelial tissue.

Lungs

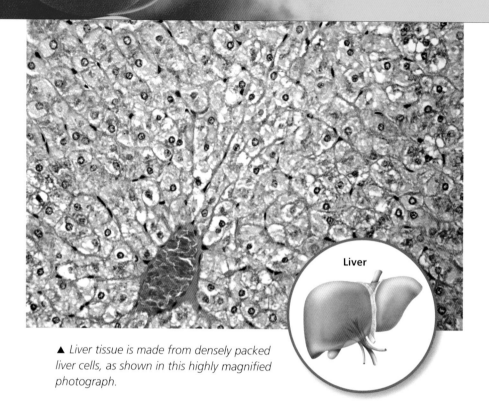

Liver

▲ *Liver tissue is made from densely packed liver cells, as shown in this highly magnified photograph.*

- **Epithelial tissue** is good lining or covering material, making skin and other parts of the body.

- **Epithelial tissue** may combine three kinds of cell to make a thin waterproof layer – squamous (flat), cuboid (box-like) and columnar (pillar-like) cells.

- **Nerve tissue** is made mostly from neurons (nerve cells), plus the Schwann cells that coat them.

- **The heart** is made mostly of muscle tissue, but also includes epithelial and connective tissue.

DID YOU KNOW?

Your body is entirely made up of tissues and fluid.

19

Organs

- **Organs are made** from combinations of tissues.

- **A collection** of related organs form a body system.

- **Body organs** include the brain, heart, lungs, kidneys and digestive organs. These all work together to keep the body functioning.

- **The largest organ** is the skin, which covers the whole body.

- **The smallest organ** is the pineal gland, a tiny organ in the brain that produces a substance that affects sleep.

- **The brain** controls the functions of many of the other body organs, making sure we keep breathing and our hearts keep beating.

- **We can survive** without some organs.

- **Some people** only have one kidney or lung or have had their appendix or spleen removed because of damage or disease.

- **Some organs**, such as the heart or liver, can be replaced by transplant surgery if they are damaged or diseased.

DID YOU KNOW?
The word 'organ' means instrument or tool in Greek.

ORGAN	WHAT IT DOES
BRAIN	Controls the nervous system
HEART	Keeps blood flowing round the body
LUNGS	Enable us to get oxygen from the air we breathe
VOICE BOX	Produces sounds that we turn into speech
STOMACH	Starts to break down food
LIVER	Produces chemicals essential for survival and digestion
GALL BLADDER	Helps digestion of food
PANCREAS	Helps to control sugar levels in the body
SPLEEN	Produces cells that help fight infections
SMALL INTESTINE	Processes food and absorbs useful substances
APPENDIX	Has no known use in humans
LARGE INTESTINE	Absorbs water from food and gets rid of unwanted material
KIDNEYS	Help control the body's fluid balance
BLADDER	Stores urine
SKIN	A protective covering over the body

Cells

- **Cells are** the basic building blocks of your body. Most are so tiny you would need 10,000 to cover a pinhead.

- **There are** over 200 different kinds of cell in your body, including nerve cells, skin cells, blood cells, bone cells, fat cells, muscle cells and many more.

- **A cell** is basically a little parcel of organic (life) chemicals with a thin membrane (casing) of protein and fat. The membrane holds the cell together, but lets nutrients in and waste out.

- **Inside the cell** is a liquid called cytoplasm, and floating in this are various minute structures called organelles.

- **At the centre** of the cell is the nucleus – this is the cell's control centre and it contains the amazing molecule DNA. This molecule not only has all the instructions the cell needs to function, but also has the pattern for new human life.

- **Each cell** is a dynamic chemical factory, and the cell's team of organelles is continually busy – ferrying chemicals to and fro, breaking up unwanted chemicals, and putting together new ones.

- **The biggest cells** in the body are nerve cells. Although the main nucleus of a nerve cell is microscopic, the tails of some cells can extend for a metre or more through the body, and be seen even without a microscope.

- **Among the smallest** cells in the body are red blood cells. These are just 0.0075 mm across and have no nucleus, since nearly their only task is ferrying oxygen.

⚡ **Most body cells** live for a very short time and are continually being replaced by new ones. The main exceptions are nerve cells – these are long-lived, and rarely replaced.

Mitochondria are the cell's power stations, turning chemical fuel supplied by the blood as glucose into energy packs of the chemical ATP

The endoplasmic reticulum is the cell's main chemical factory, where proteins are built under instruction from the nucleus

The ribosomes are the individual chemical assembly lines, where proteins are put together from basic chemicals called amino acids

The nucleus is the cell's control centre, sending out instructions via a chemical called messenger RNA whenever a new chemical is needed

The lysosomes are the cell's dustbins, breaking up any unwanted material

The Golgi bodies are the cell's despatch centre, where chemicals are bagged up inside tiny membranes to send where they are needed

▲ *This illustration shows a typical cell, and some of the different organelles (special parts of a cell) that keep it working properly. The instructions come from the nucleus in the cell's control centre, but every kind of organelle has its own task.*

Cell division

- **Cells in the body** are constantly dividing. Cells divide to replace ones that have become worn out, and during growth.

- **When a cell divides** it normally copies itself exactly.

- **The genetic material** that is in the centre of the cell splits apart and is duplicated so that the cell contains two copies.

- **The genetic material** then separates so that one copy sits in each half of the cell.

- **The cell** then gradually splits into two cells, each containing one set of the copied genetic material. This process is called mitosis.

- **Sperm and egg cells** are made by a slightly different type of cell division called meiosis. Each sperm and egg cell only has half the genetic material of other cells.

- **Occasionally** the genetic material does not copy properly and one of the new cells contains more or less genetic material than normal. This is called a mutation.

- **Mutations can be good**. Evolution takes place when good mutations give a better chance of survival and so are passed on.

- **Mutations can also** cause diseases such as cancer.

◄ *Identical twins occur when one fertilized cell splits into two.*

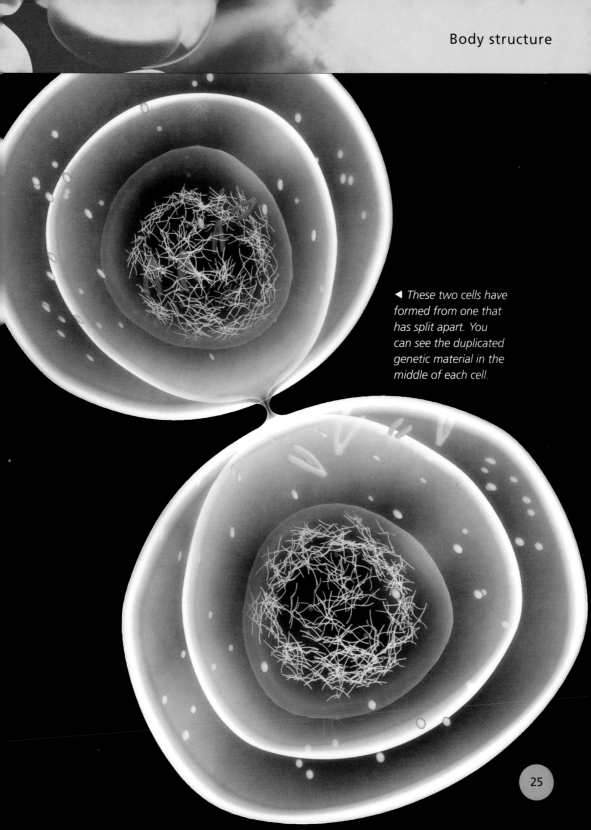

◄ These two cells have formed from one that has split apart. You can see the duplicated genetic material in the middle of each cell.

Cancer

- **Cancer is a disease** in which cells multiply abnormally, creating growths called tumours.

- **The word 'cancer'** comes from the Greek word for crab, as it spreads a bit like a crab's claw.

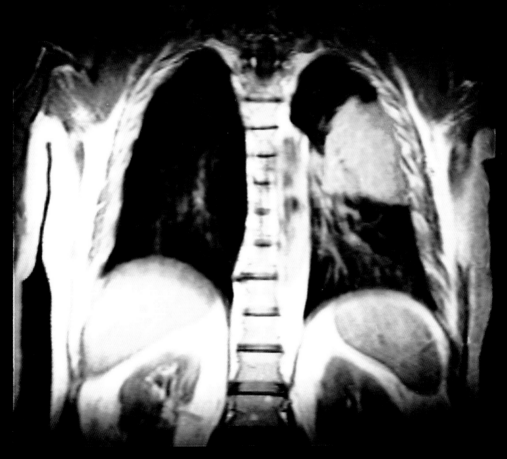

▲ *This coloured scan of a chest shows a large lung tumour (seen here as a pink area).*

DID YOU KNOW?

Cancer that has spread to another part of the body is said to have 'metastasised'.

- **Cancer can** start anywhere in the body. Once you have cancer it may spread around the body.

- **The most common** type of cancer is skin cancer. However, most skin cancers do not spread and are easily treated by removing the tumour.

- **You are much more** likely to get skin cancer if you sunbathe or use sunbeds.

- **More people** die of lung cancer than of any other type of cancer. Lung cancer is far more common in people who smoke.

- **Some types** of cancer run in families but others are caused by our lifestyle. A healthy diet with lots of vegetables and fibre can help to prevent cancer of the large intestine.

- **Infection** with some viruses increases the risk of getting certain types of cancer.

- **Cancer is usually treated** by removing the tumour. Some people also need drugs and radiation treatment to make sure all the cancer cells have been destroyed.

- **Six million people** around the world die of cancer every year. The risk increases as you get older.

▶ Some substances, such as the chemicals in cigarettes, are known to increase the risk of getting cancer.

27

Chromosomes

- **Chromosomes are** the microscopically tiny, twisted threads inside every cell that carry your body's life instructions in chemical form.

- **There are 46 chromosomes** in each of your body cells, divided into 23 pairs.

- **One of each** chromosome pair came from your mother and the other from your father.

  - **In a girl's** 23 chromosome pairs, each half exactly matches the other (the set from the mother is equivalent to the set from the father).

  - **Boys have** 22 matching chromosome pairs, but the 23rd pair is made up of two odd chromosomes.

    - **The 23rd chromosome pair** decides what sex you are, and the sex chromosomes are called X and Y.

    - **Girls have two** X chromosomes, but boys have an X and a Y chromosome.

    - **In every** matching pair, both chromosomes give your body life instructions for the same thing.

◄► A girl turns out to be a girl because she gets an X chromosome from her father. A boy gets a Y chromosome from his father.

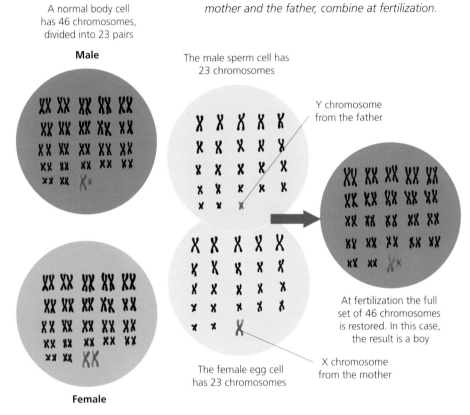

A normal body cell has 46 chromosomes, divided into 23 pairs

Male

▼ Two sets of chromosomes, one each from the mother and the father, combine at fertilization.

The male sperm cell has 23 chromosomes

Y chromosome from the father

At fertilization the full set of 46 chromosomes is restored. In this case, the result is a boy

The female egg cell has 23 chromosomes

X chromosome from the mother

Female

The chemical instructions on each chromosome come in thousands of different units called genes.

Genes for the same feature appear in the same locus (place) on each matching pair of chromosomes in every human body cell.

DNA

- **Cells are** the basic building blocks of your body. Most are so tiny you would need 10,000 to cover a pinhead.

- **DNA** (Deoxyribonucleic Acid) is the molecule inside every cell that carries all your genes. Most of the time, DNA is coiled up inside the chromosomes, but when needed, it unravels.

- **The structure** of DNA was first identified in 1953 by James Watson and Francis Crick, who announced they had 'found the secret of life'.

- **DNA is shaped** in a double helix with linking bars, like a twisted rope ladder.

- **The bars** of DNA are four special chemicals called bases – guanine, adenine, cytosine and thymine.

- **The base adenine** always pairs with thymine, and the base guanine always pairs with cytosine.

- **The bases** in DNA are arranged in groups of three called codons, and the order of the bases in each codon varies to provide a chemical code for the cell to make a particular amino acid.

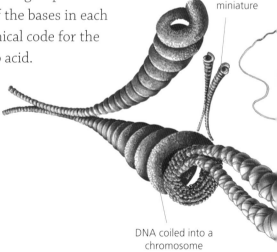

Chromosome in miniature

▶ *The sequence of bases along one strand of the DNA is a perfect mirror image of the sequence on the other side. When the strand divides down the middle, each can be used like a template to make a copy. This is how instructions are issued.*

DNA coiled into a chromosome

- **When the cell** needs to make a new protein, the DNA 'unzips' and the codons are matched by free bases, which make a copy of that part of the DNA.

- **The DNA copy** is then matched by amino acids floating in the cell, which join together in the right order to make a specific protein.

- **Because DNA** is responsible for making proteins, it is essential for growth, development and body function.

- **Each cell** contains about 2 m of DNA and if you unravelled all the DNA in your body it would stretch about 199 billion km.

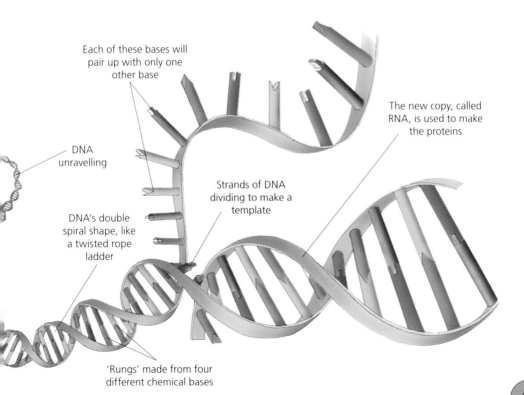

Each of these bases will pair up with only one other base

DNA unravelling

DNA's double spiral shape, like a twisted rope ladder

The new copy, called RNA, is used to make the proteins

Strands of DNA dividing to make a template

'Rungs' made from four different chemical bases

Genes

- **Genes are** the body's chemical instructions for your entire life. They hold information for growth, survival, having children and, perhaps, even for dying.

- **Individual genes** are instructions to make particular proteins – the body's building-block molecules.

- **Small sets of genes** control features such as the colour of your hair or your eyes, or create a particular body process such as digesting fat from food.

- **Each of your body cells** (except egg and sperm cells) carries identical sets of genes. This is because all your cells were made by other cells splitting in two, starting with the original egg cell in your mother.

- **Your genes** are a mixture – half come from your mother and half from your father. But none of your brothers or sisters will get the same mix, unless you are identical twins.

- **Genes make us unique** – making us tall or short, fair or dark, brilliant dancers or speakers, healthy or likely to get particular illnesses, and so on.

- **Genes are sections** of DNA – a microscopically tiny molecule inside each cell.

- **Occasionally**, genes are faulty. Some faulty genes can cause diseases and these are called genetic disorders.

DID YOU KNOW?
Each cell in your body has about 90,000 pairs of genes.

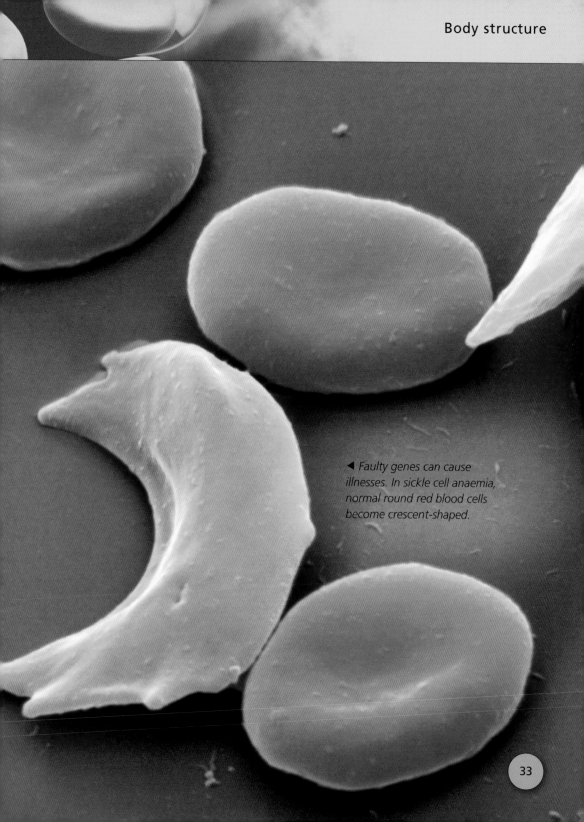

◄ *Faulty genes can cause illnesses. In sickle cell anaemia, normal round red blood cells become crescent-shaped.*

The human genome

- **The human genome** is all the genes found in the 23 pairs of chromosomes that a normal human being carries.

- **Humans have** about 20,000–25,000 genes, separated by 'junk' pieces of DNA that have no function.

- **About 97 percent** of the human genome is 'junk' and does not code for a gene.

- **The Human Genome Project** was started in 1990 and aimed to find the precise code of all the genes in the human genome. It was completed in 2003.

- **Many diseases** have a genetic component, and tests that look for a tendency to develop certain diseases, such as some types of cancer, are being developed.

▼ Just like DNA, fingerprints are unique and can be used to identify criminals.

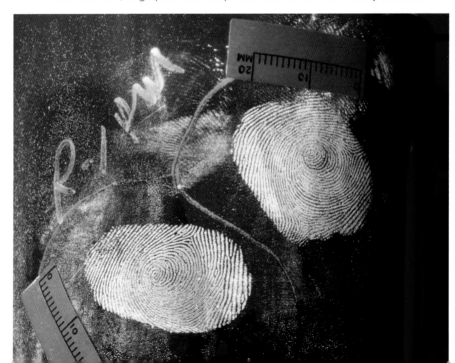

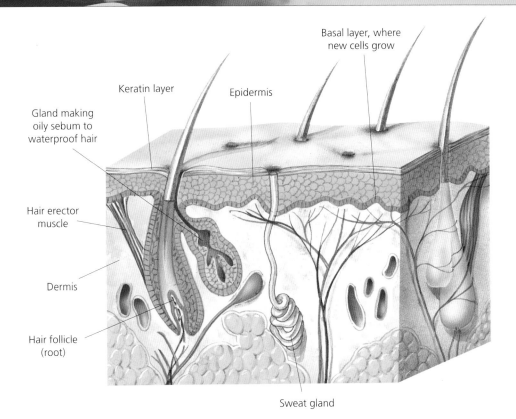

Basal layer, where new cells grow

Keratin layer

Epidermis

Gland making oily sebum to waterproof hair

Hair erector muscle

Dermis

Hair follicle (root)

Sweat gland

▲ *This is a cross-section of skin, hugely magnified, showing its key components.*

Skin is 6 mm thick on the soles of your feet, while the skin on your eyelids is just 0.5 mm thick.

Hair roots have tiny muscles that pull the hair upright when you are cold, giving you goose bumps.

The epidermis contains cells that make the dark pigment melanin – this gives dark-skinned people their colour and fair-skinned people a tan.

DID YOU KNOW?

Even though its thickness averages just 2 mm, your skin gets an eighth of all your blood supply.

39

Hair

- **Humans are** one of very few land mammals to have almost bare skin. But even humans have soft, downy hair all over, with thicker hair in places.

- **Lanugo** is the very fine hair babies are covered in when they are inside the womb, from the fourth month of pregnancy onwards.

▼ *A microscopic view of a hair. It is only alive and growing at its root, in the base of the follicle. The shaft that sticks out of the skin is dead, and is made of flat cells stuck firmly together.*

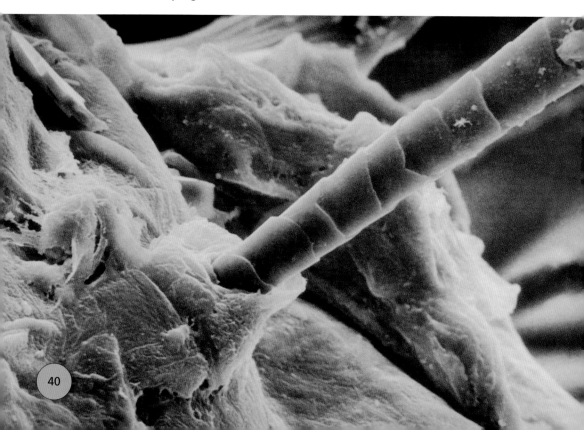

Brown hair

▶ *The colour of your hair depends upon melanin made in melanocytes at the root.*

- **Vellus hair** is fine, downy hair that grows all over your body until you reach puberty.

- **Terminal hair** is the coarser hair on your head, as well as the hair that grows on a man's chin and around an adult's genitals.

Blonde hair

- **The colour** of your hair depends on how much there are of pigments called melanin and carotene in the hairs.

- **Hair is red or auburn** if it contains carotene.

- **Black, brown and blonde hair** get its colour from black melanin.

Red hair

- **Each hair** is rooted in a pit called the hair follicle. The hair is held in place by its club-shaped tip, the bulb.

- **Hair grows** as cells fill with a material called keratin and die, and pile up inside the follicle.

- **The average person** has 120,000 head hairs and each grows about 3 mm per week.

41

Nails

- **Nails protect** the ends of our fingers and toes. They are formed of dead cells, strengthened by a protein called keratin.

- **Without nails**, we would not be able to scratch. They also help us judge pressure when picking up objects.

- **Nails grow** from a nail root, which is hidden by a fold of skin at the base of the nail called a cuticle.

- **The pale half moon** at the base of the nail is called the lunula after the Latin word for the moon.

▼ White spots on nails are very common. Many people think they are caused by a lack of calcium but they are usually due to a minor injury.

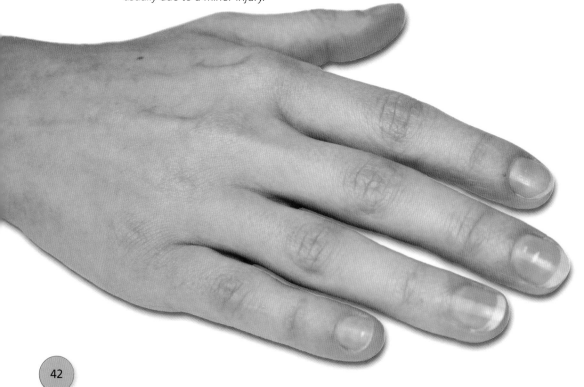

- **Most nails** only grow by about 0.5 mm a month. Fingernails grow faster than toenails.

- **Nails grow faster** in summer than in winter.

- **The nails** on your right hand will grow faster than the nails on your left hand if you are right handed. If you are left handed, the nails on your left hand will grow faster.

- **The record** for the longest nails in the world is held by a woman in the USA, who grew her nails to a total length of 8.65 m.

- **It is believed** that people have been painting their nails with nail polish or varnish since around 3000 BC.

- **Contrary to popular belief**, nails do not continue to grow after death.

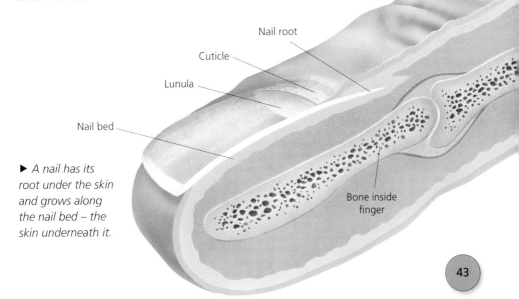

Nail root

Cuticle

Lunula

Nail bed

▶ A nail has its root under the skin and grows along the nail bed – the skin underneath it.

Bone inside finger

43

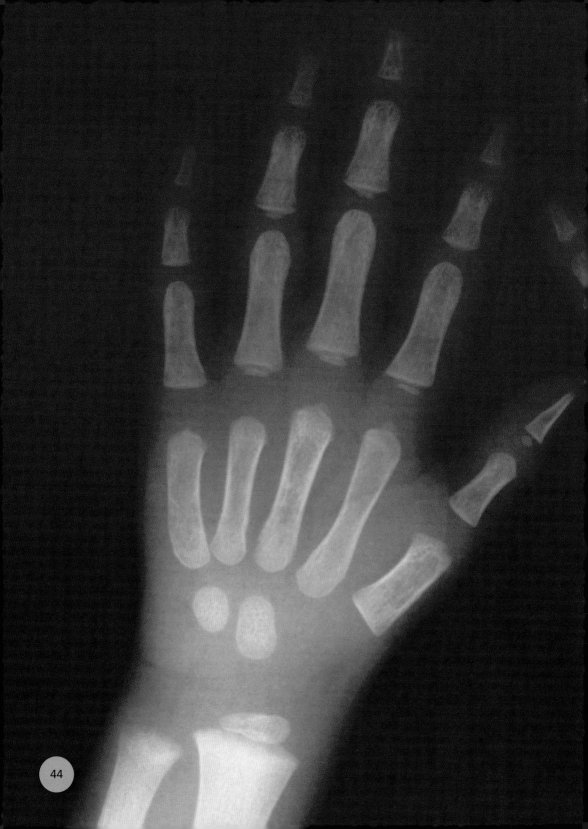

44

Musculoskeletal system

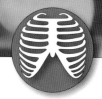

The skeleton

Your skeleton is a rigid framework of bones, which provides an anchor for your muscles, supports your skin and other organs, and protects vital organs.

An adult's skeleton has 206 bones, joined together by rubbery cartilage. Some people have extra vertebrae (the bones of the backbone, or spine).

A baby's skeleton has 300 or more bones, but some of these fuse (join) together as the baby grows.

The parts of an adult skeleton that have fused into one bone include the skull and the pelvis.

▶ *Your skeleton is the remarkably light, but very tough framework of bones that supports your body. It is made up of more than 200 bones.*

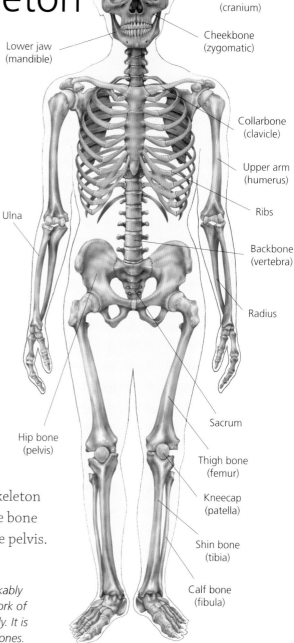

Skull (cranium)

Cheekbone (zygomatic)

Lower jaw (mandible)

Collarbone (clavicle)

Upper arm (humerus)

Ribs

Ulna

Backbone (vertebra)

Radius

Sacrum

Hip bone (pelvis)

Thigh bone (femur)

Kneecap (patella)

Shin bone (tibia)

Calf bone (fibula)

- **The skeleton** has two main parts – the axial and the appendicular skeleton.

- **The axial skeleton** is the 80 bones of the upper body. It includes the skull, the vertebrae of the backbone, the ribs and the breastbone. The arm and shoulder bones are suspended from it.

- **The appendicular skeleton** is the other 126 bones – the arm and shoulder bones, and the leg and hip bones.

- **The word 'skeleton'** comes from the ancient Greek word for dry.

- **Most women and girls** have smaller and lighter skeletons than men and boys.

▼ *Muscles pull and push against the bones on the skeleton to enable us to run and walk.*

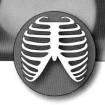

Bone

- **Bones are so strong** that they can cope with twice the squeezing pressure that granite can, or four times the stretching tension that concrete can.

- **Weight for weight**, bone is at least five times as strong as steel.

- **Bones are so light** they only make up 14 percent of your body's total weight.

- **Bones get their rigidity** from hard deposits of minerals such as calcium and phosphate.

▼ Bones are strong but very light because, on the inside, they have many holes.

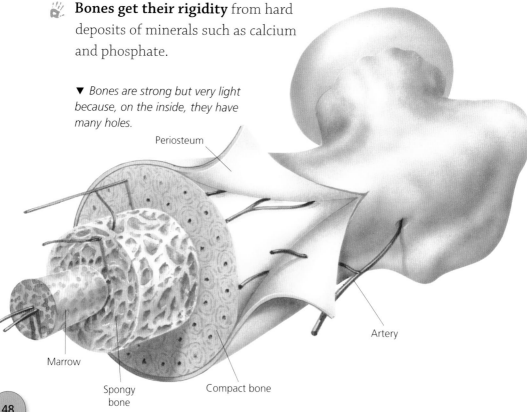

Periosteum

Marrow

Spongy bone

Compact bone

Artery

- **Bones get their flexibility** from tough, elastic, rope-like fibres of collagen.

- **The hard outside** of bones (called compact bone) is reinforced by strong rods called osteons.

- **The inside of bones** (called spongy bone) is a light honeycomb, made of thin struts or trabeculae, perfectly angled to take stress.

- **The core** of some bones, such as the long bones in an arm or leg, is called bone marrow. It is soft and jelly-like.

- **In some parts** of each bone, there are special cells called osteoblasts that make new bone. In other parts, cells called osteoclasts break up old bone.

- **Bones grow** by getting longer near the end, at a region called the epiphyseal plate.

▶ *Milk contains a mineral called calcium, which is essential for building strong bones. Babies and children need plenty of calcium to help their bones develop properly.*

49

Fractures

- **A fracture** is a broken bone.

- **Most fractures** are caused by an injury, such as a fall. Bones weakened by disease may sometimes fracture by themselves. This is called a pathological fracture.

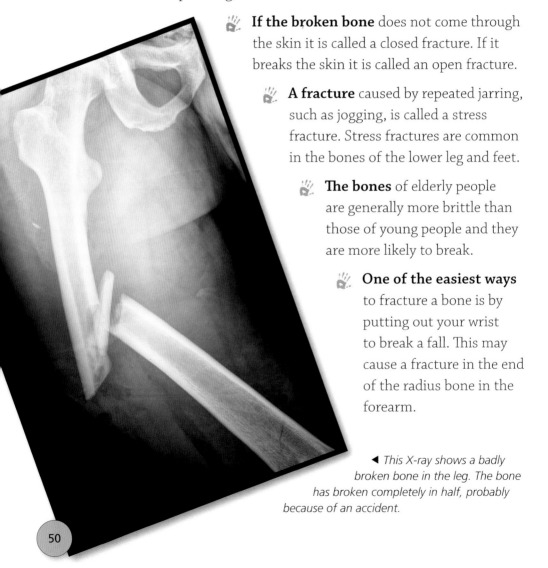

- **If the broken bone** does not come through the skin it is called a closed fracture. If it breaks the skin it is called an open fracture.

- **A fracture** caused by repeated jarring, such as jogging, is called a stress fracture. Stress fractures are common in the bones of the lower leg and feet.

- **The bones** of elderly people are generally more brittle than those of young people and they are more likely to break.

- **One of the easiest ways** to fracture a bone is by putting out your wrist to break a fall. This may cause a fracture in the end of the radius bone in the forearm.

◀ This X-ray shows a badly broken bone in the leg. The bone has broken completely in half, probably because of an accident.

▼ *Modern casts are light and waterproof but still keep broken bones still and allow them to heal.*

- **Fractures are treated** by lining up the broken ends of the bone and keeping them still in a plaster or plastic cast.

- **If a fracture** is complicated, surgery may be needed to put the bone back together properly or to hold it together with metal pins or plates.

- **Osteoblasts** (bone-making cells) heal broken bones by gradually converting tissue into bone.

- **Bones usually** take about six weeks to heal, although they may be weak for several months. Children heal faster than adults.

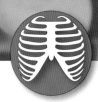

Marrow

- **Marrow** is the soft, jelly-like tissue in the middle of certain bones.

- **Bone marrow** can be red or yellow, depending on whether it has more blood tissue or fat tissue.

- **Red bone marrow** is the body's blood factory. This is where all blood cells, apart from some white cells, are made.

- **All bone marrow** is red when you are a baby, but as you grow older, more and more turns yellow.

- **In adults**, red marrow is only found in the ends of the limbs' long bones, breastbone, backbone, ribs, shoulder blades, pelvis and the skull.

- **Yellow bone marrow** is a store for fat, but it may turn to red marrow when you are ill.

- **All the different kinds** of blood cell start life in red marrow as one type of cell called a stem cell. Different blood cells then develop as the stem cells divide and re-divide.

- **Some stem cells** divide to form red blood cells and platelets.

- **Other stem cells** divide to form lymphoblasts. These divide in turn to form various different kinds of white cells, such as monocytes and lymphocytes.

- **The white cells** made in bone marrow play a key part in the body's immune system. This is why bone-marrow transplants can help people with illnesses that affect their immune system.

> **DID YOU KNOW?**
> Animal bone marrow is an important food in many cultures.

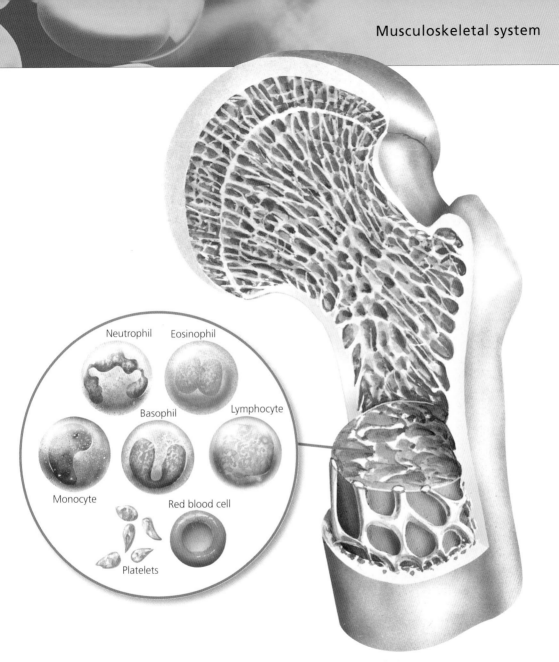

Neutrophil   Eosinophil

Basophil   Lymphocyte

Monocyte

Red blood cell

Platelets

▲ Inside the tough casing of most bones is a soft, jelly-like core called the marrow, which can be either red or yellow. The red marrow in certain bones is the body's blood-cell factory, making five million new cells a day. Some varieties of blood cells are shown above.

The skull

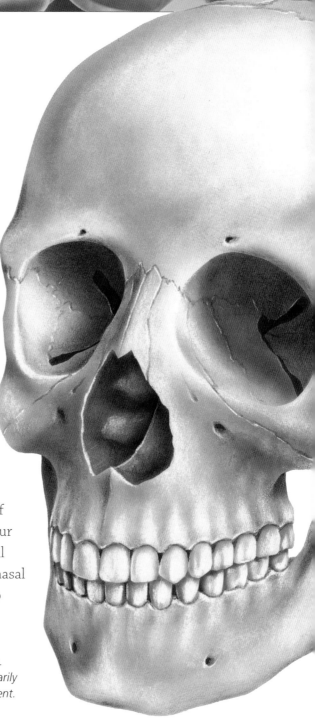

- **The skull** or cranium is the hard bone case that contains and protects the brain.

- **Although it looks** as though the skull is a single bone, it is actually made up of 22 separate bones, cemented together along rigid joints called sutures.

- **The dome** on top is called the cranial vault and it is made from eight curved pieces of bone fused (joined) together.

- **As well as the sinuses** of the nose, the skull has four large cavities – the cranial cavity for the brain, the nasal cavity (the nose) and two orbits for the eyes.

▶ *Skulls vary in size and shape. A bigger skull does not necessarily mean a person is more intelligent.*

▶ A child's skull, shown here in this X-ray photo, is quite large in relation to the rest of the child's body. As our bodies grow, our skull starts to look smaller in proportion.

There are holes in the skull to allow blood vessels and nerves through, including the optic nerves to the eyes and the olfactory tracts to the nose.

The biggest hole is in the base. It is called the foramen magnum, and the brain stem goes through it to meet the spinal cord.

In the 19th century, people called phrenologists thought they could work out people's characters from little bumps on their skulls.

Archaeologists can reconstruct faces from the past using computer analysis of ancient skulls.

55

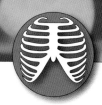

Sinuses

The sinuses are air-filled holes in the skull around the eyes and the nose.

Sinuses develop gradually as you grow and help make the skull lighter.

The sinuses also make the voice resonate and may help to protect the face against blows by acting a little like an air bag.

Most sinuses lie in pairs – one on each side of the face.

There are two sinuses in the forehead (frontal sinuses), several by the sides of the nose (ethmoid sinuses), one in each of the cheekbones (maxillary sinuses) and two deep in the skull (sphenoid sinuses).

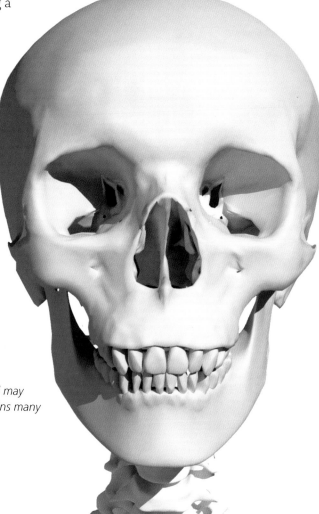

▶ Although the human skull may look like solid bone, it contains many spaces called sinuses.

▲ *Most of the sinuses lie in the bones around the nose and cheeks.*

DID YOU KNOW?
Scientific studies have shown that some dinosaurs had sinuses.

The sinuses are lined with glands that produce mucus, which passes through small holes in the bones of the skull and into the back of the nose.

The mucus traps dust and tiny particles and helps to keep the air in your airways moist and warm.

The linings of the sinuses may become swollen when you have a cold or if you have an allergy. This causes a headache and is called sinusitis.

Sinusitis is very common and most people will get it at some point in their lifetime.

The word 'sinus' is Latin and means a fold or pocket.

The backbone

DID YOU KNOW?

Gravity squashes the joints in the spine over the course of a day, so you are shorter in the evening than you are in the morning.

The backbone, otherwise known as the spine, extends from the base of the skull down to the hips.

It is not a single bone, but a column of drum-shaped bones called vertebrae (singular, vertebra).

There are 33 vertebrae in total, although some of these fuse or join as the body grows.

Each vertebra is linked to the next by small facet joints, which give limited movement.

The vertebrae are separated by discs of rubbery material called cartilage. These cushion the bones when you run and jump.

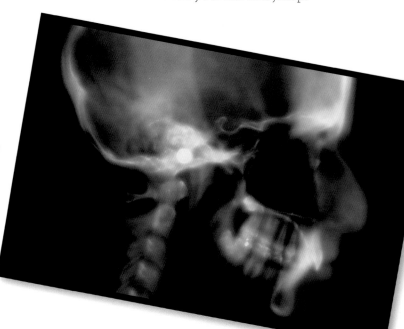

► The joint between the top two bones of the spine allows us to turn our heads.

58

The bones of the spine are divided into five groups from top to bottom. These are the cervical (7 bones), the thoracic (12 bones), the lumbar (5 bones), the sacrum (5 bones fused together), and the coccyx (4 bones fused together).

The cervical spine is the vertebrae of the neck. The thoracic spine is the back of the chest, and each bone has a pair of ribs attached to it. The lumbar spine is the small of the back.

A normal spine curves in an S-shape, with the cervical spine curving forwards, the thoracic section curving backwards, the lumbar forwards, and the sacrum curving backwards.

On the back of each vertebra is a bridge called the spinal process. The bridges on each bone link together to form a tube that holds the spinal cord, the body's central bundle of nerves.

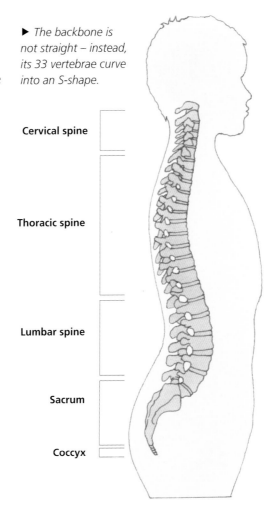

▶ The backbone is not straight – instead, its 33 vertebrae curve into an S-shape.

Cervical spine

Thoracic spine

Lumbar spine

Sacrum

Coccyx

Ribs

▼ *The ribs provide a framework for the chest and form a protective cage around the heart, lungs and other organs.*

1 Lung

2 Sternum (breastbone)

3 Heart

4 Stomach

5 False ribs

6 Liver

7 True ribs

8 Costal cartilage

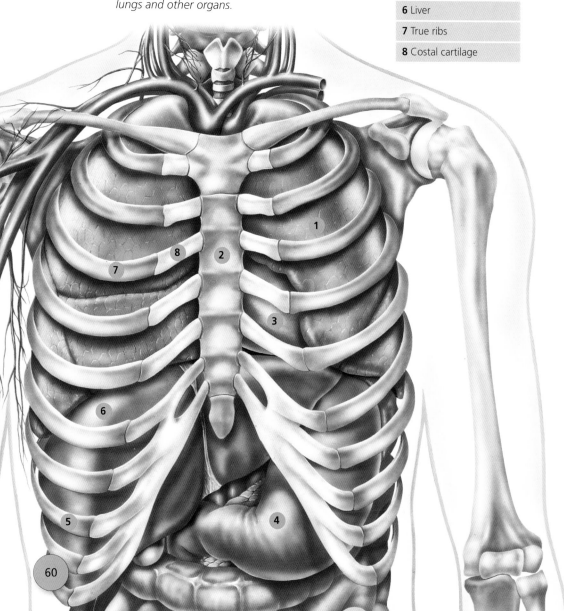

- **The ribs** are the thin, flattish bones that curve around your chest.

- **Together**, the rib bones make up the rib cage.

- **The rib cage** protects the backbone and breastbone, as well as your vital organs – heart, lungs, liver, kidneys, stomach, spleen and so on.

DID YOU KNOW?
Some people don't have all their ribs, whereas other people have an extra pair.

- **You have 12 pairs** of ribs altogether.

- **Seven pairs** are called true ribs. Each rib is attached to the breastbone in front and curves around to join on to one of the vertebrae that make up the backbone via a strip of costal cartilage.

- **There are three pairs** of false ribs. These are attached to vertebrae but are not linked to the breastbone. Instead, each rib is attached to the rib above it by cartilage.

- **There are two pairs** of floating ribs. These are attached only to the vertebrae of the backbone.

- **The gaps** between the ribs are called intercostal spaces, and they contain thin sheets of muscle that expand and relax the chest during breathing.

- **Flail chest** is when many ribs are broken (often in a car accident) and the lungs heave the chest in and out.

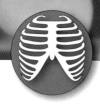

The pelvis

- **The pelvis** is made up of two hip bones, each of which is made up of three bones that fuse together at puberty – the ilium, the pubis and the ischium.

- **The pelvic bones** join the sacrum (part of the backbone) at the back and are fused together by cartilage at the front.

- **Organs** in the lower part of the abdomen, such as the bladder and female reproductive organs, are protected by the pelvis.

- **The pelvis also supports** the weight of the upper body and helps transfer weight to and from the legs when standing and walking.

- **Strong muscles** in the thigh and buttocks enable us to walk and sit. These are attached to the pelvis.

- **The human pelvis** is shaped differently to every other animal's to allow us to walk upright easily.

▼ *The pelvic bones support and protect the organs of the pelvis. This illustration shows a female pelvis containing the reproductive organs (1) and the bladder (2).*

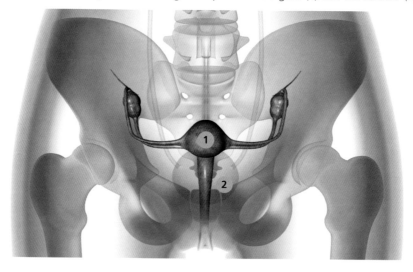

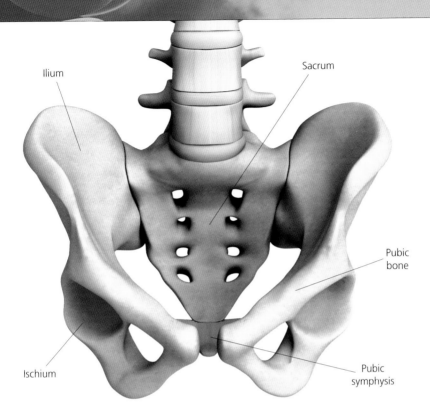

Ilium

Sacrum

Pubic bone

Ischium

Pubic symphysis

▲ *The two pelvic bones are connected to the lower back of the backbone.*

In women and girls, the pelvis is much wider than in men and boys. This is because the opening has to be wide enough for a baby to pass through when it is born.

Forensic scientists are able to use this difference in pelvis shape to tell whether a skeleton belonged to a man or a woman.

During labour, hormones cause the cartilage at the front of the pelvis to loosen, allowing the baby to pass through the pelvis more easily.

63

Joints

- **Body joints** are places where bones meet. There are about 400 joints in the human body.

- **Every time** you sit down, walk, play a sport or move at all you are using your joints.

- **Joints provide flexibility** by allowing movement and create stability by holding bones together.

- **Most body joints** let bones move, but some kinds of joint only let them move in certain ways.

- **Only fixed joints**, such as those in the skull, allow no movement.

- **All other joints** contain cartilage to cushion the ends of the bones. Some joints also contain fluid to lubricate the bones.

▶ Gymnasts must have supple, flexible joints in order to perform complicated exercises such as this.

- **Semi-movable joints**, such as those in the pelvis and backbone, allow a little movement.

- **Synovial joints** are flexible joints, lubricated with oily 'synovial fluid' and cushioned by cartilage.

- **A dislocated joint** occurs when bones are forced out of place, often by playing sport.

- **The cartilage** in joints may wear out through age or overuse through sport. In some cases surgery may be needed to replace the joint with an artificial one.

DID YOU KNOW?
You have about 30 small joints in each hand and wrist.

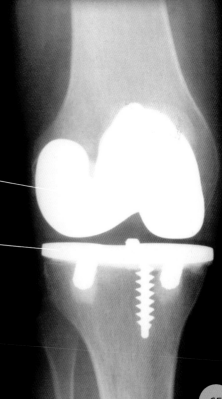

Cap on base of thigh bone

Plate screwed into top of shin bone

▶ This X-ray shows an artificial knee joint. The ends of the two bones have been replaced with metal surfaces.

Fixed and semi-movable joints

- **In a fixed joint**, bones are held together by fibrous tissue so that they do not move or only move very slightly.

- **The skull** is not one bone, but 22 separate bones bound tightly together with fibres so that they cannot move.

- **These fixed joints** in the skull are called sutures.

- **The skull bones** are not fused together at birth. The soft spots where the bones are not joined are called fontanelles and usually disappear by about 18 months.

▶ *The jagged lines in this photograph show where the bones of the skull have fused together.*

Suture

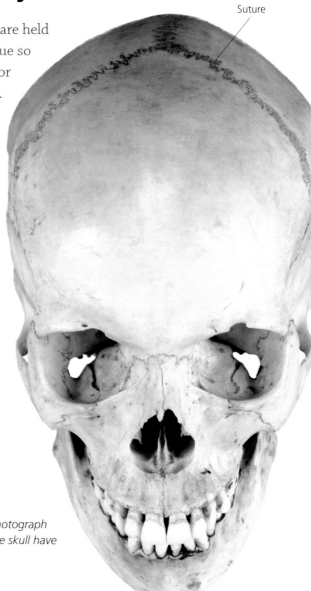

- **In a semi-movable joint**, the bones are held together by cartilage that only allows a little movement.

- **The pubic symphysis** is a semi-movable joint that holds the two pelvic bones together at the front of the pelvis.

- **In women**, the pubic symphysis loosens during labour to allow the baby to be born more easily.

- **There are semi-movable joints** between each of the bones (vertebrae) of the backbone. Although each only moves slightly, together they provide flexibility.

- **These relatively inflexible joints** in the spine are cushioned by pads of cartilage.

- **The cartilage** acts as shock absorbers for the spine when we run or jog.

DID YOU KNOW?

The cartilage in the joints in your spine is stronger than in other joints.

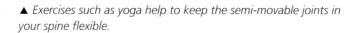

▲ *Exercises such as yoga help to keep the semi-movable joints in your spine flexible.*

67

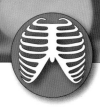

Synovial joints

- **Synovial joints** are freely movable joints and are the most common type of joint in the body.

- **In a synovial joint** the bones are cushioned with cartilage and the joint space is filled with a lubricating fluid called synovial fluid. The joint is encased by a capsule that is often reinforced by ligaments that hold it all together.

- **There are six types** of synovial joint.

- **In ball-and-socket joints**, such as the shoulder and hip, the rounded end of one bone sits in the cup-shaped socket of the other, and can move in almost any direction.

- **Hinge joints**, such as the elbow, knees, fingers and toes, let the bones swing to and fro in two directions like door hinges.

- **Swivel joints** turn like a wheel on an axle. Your head can swivel to the left or to the right on your spine.

- **Saddle joints** such as those in the thumb have the bones interlocking like two saddles. These joints allow great mobility with considerable strength.

- **Plane joints** in the wrist and the foot allow almost flat bone surfaces to slide over each other but only in short distances.

- **In ellipsoidal joints** the domed end of one bone fits into a cavity on the other bone allowing a limited rotation. These types of joints are found at the base of the fingers and in the wrist.

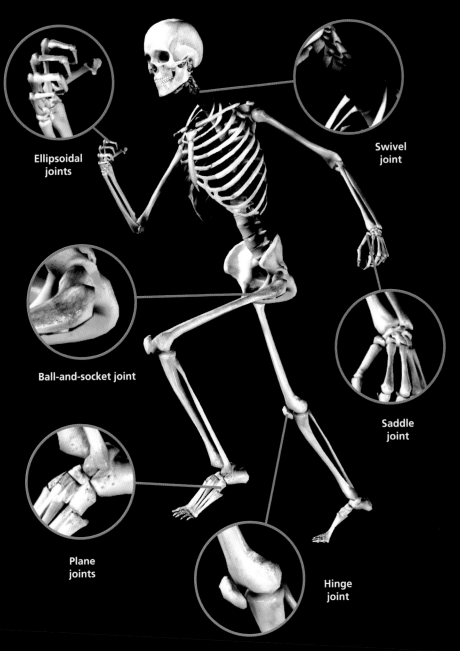

Ellipsoidal
joints

Swivel
joint

Ball-and-socket joint

Saddle
joint

Plane
joints

Hinge
joint

▲ *Synovial joints allow the body to move in many ways so we can walk,*
run, play and work.

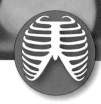

Knee joint

- **The knee** is a synovial joint and the biggest joint in the body.

- **It acts as a hinge** between the femur (thighbone) and the two bones in the lower leg.

- **The knee joint** doesn't just straighten or bend, it also rotates slightly.

- **Cartilage in the knee** makes two dish shapes called menisci (singular, meniscus) between the thigh and lower leg bones.

- **These cartilages** are often damaged when playing sport such as football or rugby, especially if the leg twists suddenly with the knee bent.

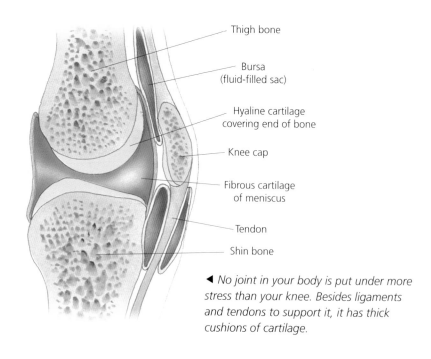

Thigh bone

Bursa (fluid-filled sac)

Hyaline cartilage covering end of bone

Knee cap

Fibrous cartilage of meniscus

Tendon

Shin bone

◀ *No joint in your body is put under more stress than your knee. Besides ligaments and tendons to support it, it has thick cushions of cartilage.*

▲ *Falling at an awkward angle, which can happen during sports such as rugby, may cause a knee injury.*

- **Cartilage can also** become damaged with age or through overuse, especially in people who do a lot of squatting or lifting, which puts extra pressure on the knee.

- **The knee** is constantly under stress when getting up or sitting down, walking or running.

- **For the knee joint** to work well, the joint needs to be in good condition and the muscles around the knee need to be strong.

- **The knee** is one of the joints in the body that can be replaced by an artificial joint if it becomes too damaged or diseased to work properly.

- **The knee cap** is a small bone that sits outside the joint and is held in place by tendons that attach to the muscles of the upper and lower legs.

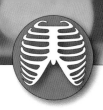

Cartilage

Cartilage is a rubbery substance used in various places around the body. You can feel cartilage in your ear if you move it back and forward.

▼ *A single blow to the nose can easily damage the nasal cartilage, as often happens to boxers.*

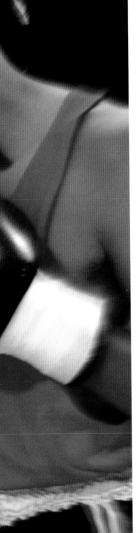

Cartilage is made from cells called chondrocytes embedded in a jelly-like substance with fibres of collagen, all wrapped in an envelope of tough fibres.

There are three types of cartilage: hyaline, fibrous and elastic.

Hyaline cartilage is the most widespread in your body. It is almost clear, pearly white and quite stiff. Hyaline cartilage is used in many of the joints between bones to cushion them against impacts.

Fibrous cartilage is really tough cartilage used in between the bones of the spine and in the knee.

Elastic cartilage is very flexible and used in your airways, nose and ears.

Cartilage grows quicker than bone, and the skeletons of babies in the womb are mostly cartilage, which gradually ossifies (hardens to bone).

Osteoarthritis is when joint cartilage breaks down, making certain movements painful.

Scientists can now make artificial cartilage that can be used to help repair damaged joints.

73

Tendons and ligaments

✋ **Tendons are cords** that tie a muscle to a bone or tie a muscle to another muscle.

✋ **Most tendons** are round, rope-like bundles of fibre. A few, such as the ones in the abdomen wall, are flat sheets called aponeuroses.

✋ **Tendon fibres** are made from a rubbery substance called collagen.

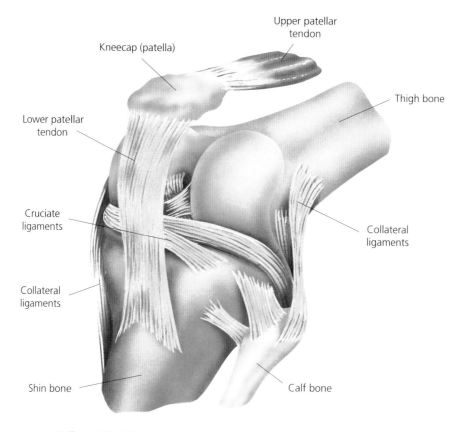

Upper patellar tendon

Kneecap (patella)

Thigh bone

Lower patellar tendon

Cruciate ligaments

Collateral ligaments

Collateral ligaments

Shin bone

Calf bone

▲ Collateral (side) ligaments stop the knee wobbling from side to side. Cruciate (crossing) ligaments tie the knee across the middle to stop it bending or straightening too much. Tendons hold the kneecap in place.

- **Your fingers** are moved mainly by muscles in the forearm, which are connected to the fingers by long tendons.

- **The Achilles tendon** pulls up your heel at the back.

- **Ligaments are cords** attached to bones on either side of a joint. They strengthen the joint.

- **Ligaments also** support various organs, including the liver, bladder and uterus (womb).

- **Women's breasts** are held in place by bundles of ligaments.

- **Ligaments are made** up of bundles of tough collagen and a stretchy substance called elastin.

▶ *Tendons provide a link between muscle and bone. They prevent muscles tearing when they are put under strain.*

75

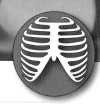

Muscles

DID YOU KNOW?
Your tongue is made from 16 muscles.

- **Muscles are special fibres** that contract (tighten) and relax to move parts of the body.

- **Muscles give our bodies** their different shapes and help to hold the body upright.

- **Most muscles** attach to bones using tendons and most muscles cross over a joint so they can move the joint.

- **Muscles are usually** arranged in pairs, because although muscles can shorten themselves, they cannot forcibly make themselves longer. So the flexor muscle that bends a joint is paired with an extensor muscle to straighten it out again.

- **There are three types** of muscle – skeletal, smooth and heart muscle.

- **Most of the muscle** in the body is skeletal muscle. There are around 640 skeletal muscles and they make up about half the weight of your body.

- **Your body's longest muscle** is the sartorius, on the inner thigh.

- **The widest muscle** is the external oblique, which runs around the side of the upper body.

- **Your body's biggest muscle** is the gluteus maximus in your buttock (bottom).

- **The shortest muscle** is the stapedius, which attaches to one of the tiny bones in the ear.

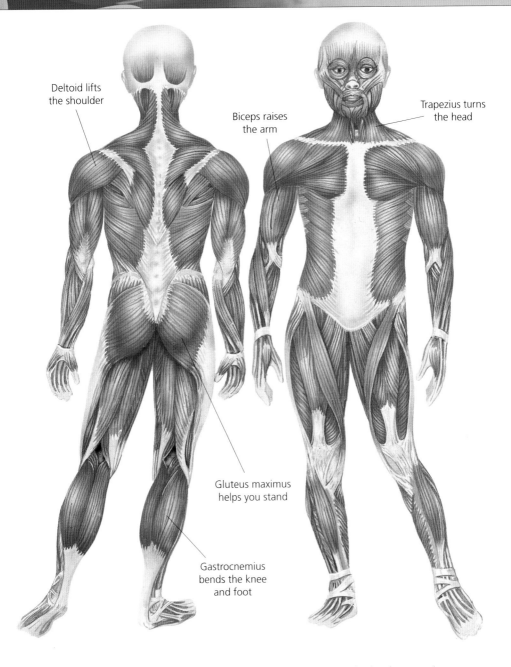

Deltoid lifts
the shoulder

Biceps raises
the arm

Trapezius turns
the head

Gluteus maximus
helps you stand

Gastrocnemius
bends the knee
and foot

▲ *Your body has several layers of muscle. Most are attached to bones using tough fibres called tendons.*

Muscle types

- **Voluntary muscles** are all the muscles you can control by will (conscious thought), such as your arm muscles.

- **Involuntary muscles** are the muscles you cannot control at will, but which work automatically, such as the muscles that move food through your intestine.

- **Most voluntary muscles** cover the skeleton and are therefore called skeletal muscles.

- **Skeletal muscles** are made of special cells called myofibrils.

Body of muscle

- **Hundreds or thousands** of these fibres bind together like fibres in string to form each muscle.

- **Each myofibril** is marked with dark bands, giving the muscle its name of stripy or 'striated' muscle.

- **Skeletal muscle** is strong but cannot keep working for long periods of time like other types of muscle. Athletes have to train so that their skeletal muscles will work for longer.

- **Most involuntary muscles** form sacs or tubes such as the intestine or the blood vessels. They are called smooth muscle because they lack the bands or stripes of voluntary muscles.

- **Heart muscle** is a unique combination of skeletal and smooth muscle. It has its own built-in contraction rhythm of 70 beats a minute, and special muscle cells that work like nerve cells for transmitting the signals for waves of muscle contraction to sweep through the heart.

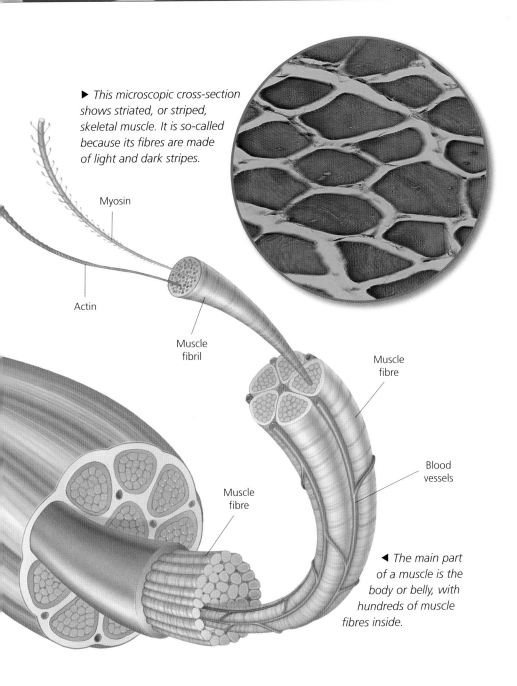

▶ This microscopic cross-section shows striated, or striped, skeletal muscle. It is so-called because its fibres are made of light and dark stripes.

Myosin

Actin

Muscle fibril

Muscle fibre

Blood vessels

Muscle fibre

◀ The main part of a muscle is the body or belly, with hundreds of muscle fibres inside.

Muscle movement

▪ **Most muscles** are long and thin and they work by pulling themselves shorter – sometimes contracting by up to half their length in response to signals from the brain.

▪ **The stripes** in skeletal muscle are alternate bands of filaments of two substances: actin and myosin.

▪ **The actin and myosin** interlock, like teeth on a zip.

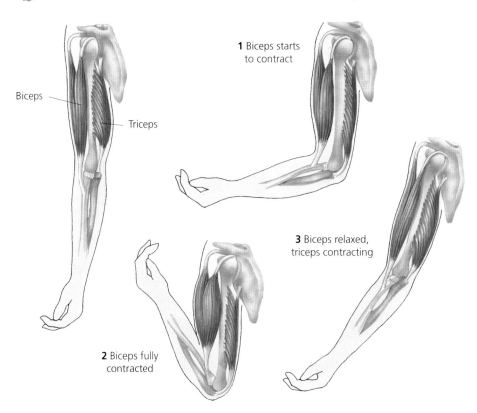

Biceps

Triceps

1 Biceps starts to contract

3 Biceps relaxed, triceps contracting

2 Biceps fully contracted

▲ *Muscles, such as the biceps and triceps in the upper arm, work in pairs, pulling in opposite directions to one another.*

When a nerve signal comes from the brain, chemical 'hooks' on the myosin twist, yanking the actin filaments along, and shortening the muscle.

The chemical hooks on myosin are made from a stem called a cross-bridge and a head made of a chemical called adenosine triphosphate or ATP.

ATP is sensitive to calcium, and the nerve signal from the brain that tells the muscle to contract does its work by releasing a flood of calcium to trigger the ATP.

Muscles can only pull and not push and are usually paired across joints.

When one muscle contracts it pulls on the joint, causing movement. The muscle on the other side of the joint will be relaxed.

To move the joint back to its original position the contracted muscle relaxes and the opposing muscle contracts.

▶ Training with weights strengthens individual muscles, such as the biceps in the arm.

81

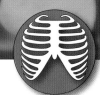

The arm

- **Three long bones**, linked by a hinge joint at the elbow, make up the arm.

- **The two bones** of the lower arm are the radius and the ulna.

- **The radius** supports the thumb side of the wrist; the ulnar supports the outside of the wrist.

- **Major arteries** come nearer the surface at the wrist than at almost any other place in the body, so the wrist is one of the best places to test the pulse.

Humerus

Ulna

Radius

Ulnar artery

Radial artery

- **The bone** of the upper arm is the humerus. It connects to the shoulder blade (the scapula) at the shoulder joint, which is a ball-and-socket joint.

- **There are two major muscles** of the upper arm – the biceps (which bends the elbow), and the triceps (which straightens it).

- **The muscles** in the lower arm help to turn and move the wrist, hand and fingers.

- **The shoulder** is one of the most flexible but least stable joints of the skeleton, since it is set in a very shallow socket. Dislocated shoulders, in which the bone is dislodged from the joint, are common in contact sports such as rugby.

◄ Look at the inside of your wrist on a warm day and you may be able to see the radial artery beneath the skin.

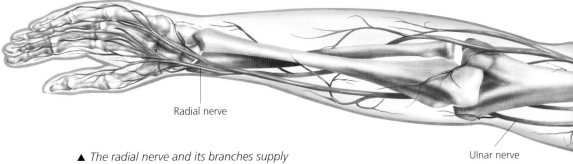

Radial nerve

Ulnar nerve

▲ *The radial nerve and its branches supply the dorsal muscles (the muscles of the back and shoulder).*

Shoulder joints are supported by six major muscle groups, including the powerful deltoid (shoulder) muscles.

Because of the wide range of movement of the shoulder and the ability to rotate the forearm and wrist, we can use our arms to do many different activities.

DID YOU KNOW?

The term 'funny bone' refers to the upper arm bone or humerus – a pun on the word 'humorous'.

▼ *A complicated network of muscles and nerves allow you to move your arm in many directions.*

Triceps straightens arm

Extensor muscles straighten fingers

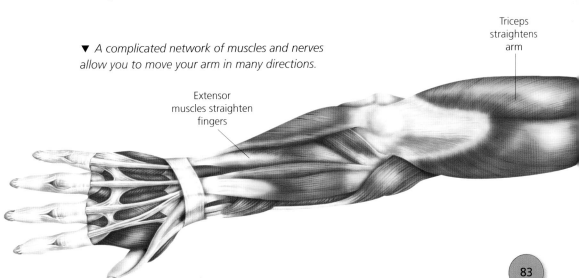

The hand

- **The hand** is made from 26 bones, including the carpals (wrist bones), the metacarpals (hand bones) and the phalanges (finger bones).

- **The wrist** contains eight small bones called the carpals.

- **There are no** strong muscles in the hand. When you grip firmly, most of the power comes from muscles in the lower arm, linked to the bones of the hand by long tendons.

- **Humans and other primates**, such as monkeys, are different to all other animals as we have opposable thumbs. This means our thumbs can touch the tips of the fingers on the same hand.

- **There are more** nerve endings in the hand than anywhere else in the body. The fingertips are especially sensitive to touch.

◄ The intricate network of bones in your hands enables you to perform delicate and complex movements such as writing or playing a musical instrument.

▲ *Muscles and nerves in your hand and lower arm all act together with your brain so you can use equipment like mobile phones.*

Between 70 and 80 percent of people are right-handed, which means that they have more co-ordination in their right hand than in their left.

In the past, children who were left-handed were forced to write with their right hand. Even today, most tools, instruments and devices are made for right-handed people.

Some people have more than the usual numbers of fingers on one or both hands. This is called polydactyly.

We use our hands all the time. They are capable of delicate gestures, such as stroking a cat, or tough activities such as digging the garden. We can play instruments or make beautiful objects and even use them to communicate through sign language.

The leg

- **The leg** is made from three long bones, linked by a hinge joint at the knee.

- **The femur** (thigh bone) is the body's longest bone.

- **It connects** to the pelvis at the hip joint, which is a ball-and-socket joint.

- **The hip** is one of the strongest joints in the body. The ball at the end of the femur sits in a very deep cup in the pelvis, making it almost impossible to dislocate.

- **The two bones** of the lower leg are the tibia and fibula.

- **The tibia** is the main shinbone at the front of the leg; the fibula lies to the outer side and slightly behind.

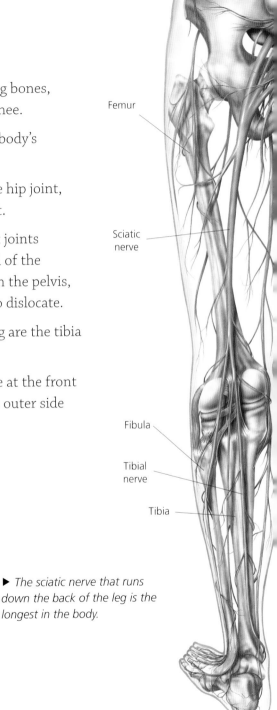

Femur

Sciatic nerve

Fibula

Tibial nerve

Tibia

DID YOU KNOW?

In anatomy, the word 'leg' is just used for the part of the leg between the knee and ankle.

▶ The sciatic nerve that runs down the back of the leg is the longest in the body.

- **The muscles** of the leg are especially strong to enable us to sit, stand, walk and run. The largest muscle in the body, the gluteus maximus, is used for stepping or standing up.

- **Muscles in the lower leg** help to turn and move the ankle, foot and toes.

- **Human legs** are longer in proportion to their bodies than those of most other animals. This is because we have adapted to walk upright on two legs.

- **Since the start** of the 1900s, women in Western cultures have been removing hair from their legs as a sign of beauty.

▲ *Our legs contain large bones, muscles and nerves so they can take the weight and strain of our bodies.*

87

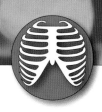

The foot

- **The foot** is made from 24 bones, including the tarsus (ankle bones), the metatarsus (foot bones) and the phalanges (toe bones).

- **The ankle** and part of the foot nearest the ankle contains eight bones called the tarsus bones.

- **Movement of the ankle**, foot and toes mostly comes from muscles in the lower leg, linked to the bones of the foot by long tendons.

- **We use our feet** and toes for balance. Our big toes are especially important; without them it would be much harder to walk and stand upright.

- **Feet usually** have an arch in the middle formed by the shape of the bones and strong tendons. Someone with no arch is said to have 'flat feet'.

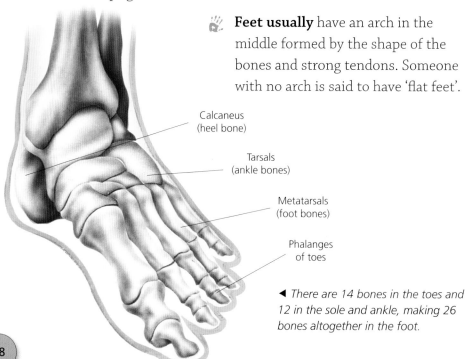

Calcaneus (heel bone)

Tarsals (ankle bones)

Metatarsals (foot bones)

Phalanges of toes

◀ There are 14 bones in the toes and 12 in the sole and ankle, making 26 bones altogether in the foot.

◀ *Ballerinas who dance en pointe (on their toes) have to wear special shoes to help support their feet.*

There are many customs associated with feet around the world. In some countries it is considered offensive to expose the soles of your feet to another person. It is also thought rude to touch with your feet or point your feet towards someone else.

In many countries it is thought polite to remove your shoes when you enter someone else's home.

In China, young girls used to have their feet bound to make them very small. This was thought to be attractive but was very painful and is now no longer done.

The world record for the most fingers and toes in one person is 25, with 12 fingers and 13 toes.

89

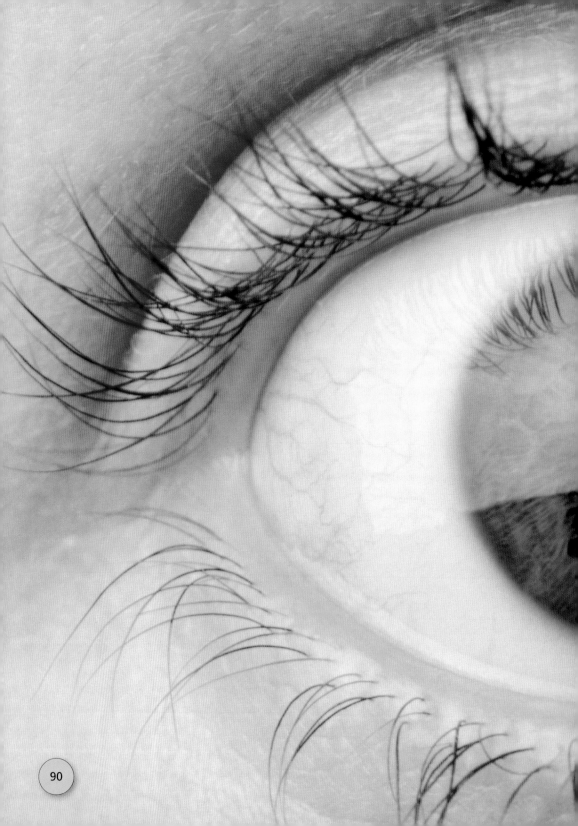

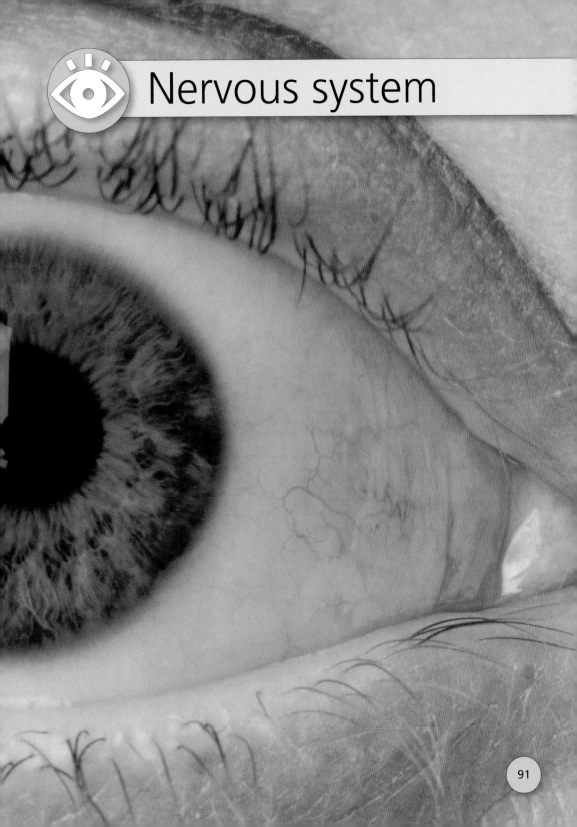

Nervous system

The nervous system

- **The nervous system** is your body's control and communication system, made up of nerves and the brain.

- **Nerves are** your body's hot lines, carrying instant messages from the brain to every organ and muscle – and sending back an endless stream of data to the brain about what is going on both inside and outside your body. The nervous system is divided into two parts – the central nervous system and the peripheral nervous system.

- **The central nervous system** (CNS) consists of the brain and spinal cord.

- **The peripheral nervous system** (PNS) is made up of the nerves that branch out from the CNS to the rest of the body.

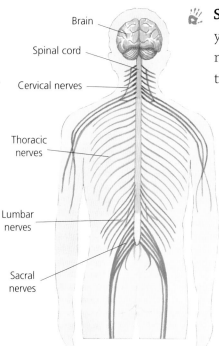

Brain

Spinal cord

Cervical nerves

Thoracic nerves

Lumbar nerves

Sacral nerves

- **Some PNS nerves** are as wide as your thumb. The longest is the sciatic nerve, which runs from the base of the spine to the knee.

- **The PNS** can be divided into nerves that control voluntary actions, such as walking or throwing a ball, and autonomic nerves that control all the body's functions.

◀ Spinal nerves branch off the spinal cord in pairs, with one nerve on either side. They are arranged in four groups, and there is one pair between each of the 32 vertebrae.

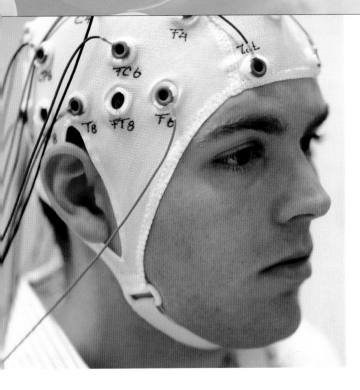

◄ *The process of having brain waves measured is called electroencephalography. It is used to help diagnose nervous system disorders.*

The autonomic nervous system (ANS) is part of the PNS. It controls all internal body processes such as breathing automatically, without you even being aware of it.

The ANS is split into two complementary (balancing) parts – the sympathetic and the parasympathetic. The sympathetic system speeds up body processes when they need to be more active, such as when the body is exercising or under stress. The parasympathetic slows them down.

Nerves can be divided into motor nerves, which carry messages from the brain to muscles to control movement, and sensory nerves, which carry messages from sensory receptors in the body to the brain.

In many places, sensory nerves run alongside motor nerves.

Central nervous system

- **The central nervous system** (CNS) is made up of the brain and the spinal cord (the nerves of the spine).

- **It is responsible** for collecting information from all the other nerves in the body, processing data and sending out appropriate responses.

- **The CNS** contains billions of densely packed interneurons – nerve cells with very short connecting axons (tails).

- **A surrounding bath of liquid** called cerebrospinal fluid cushions the CNS from damage.

- **There are 86 main nerves** branching off the CNS.

- **There are 12 pairs** of cranial nerves and 31 pairs of spinal nerves.

- **Cranial nerves** are the 12 pairs of nerves that branch off the CNS out of the brain.

- **Spinal nerves** are the 31 pairs of nerves that branch off the spinal cord.

- **The spinal nerves** are made up of eight cervical nerve pairs, 12 thoracic pairs, five lumbar pairs, five sacral pairs and one coccyx pair.

- **Many spinal nerves** join up just outside the spine in five 'spaghetti junctions' called plexuses.

> **DID YOU KNOW?**
> The CNS sends out messages to more than 640 muscles around the body.

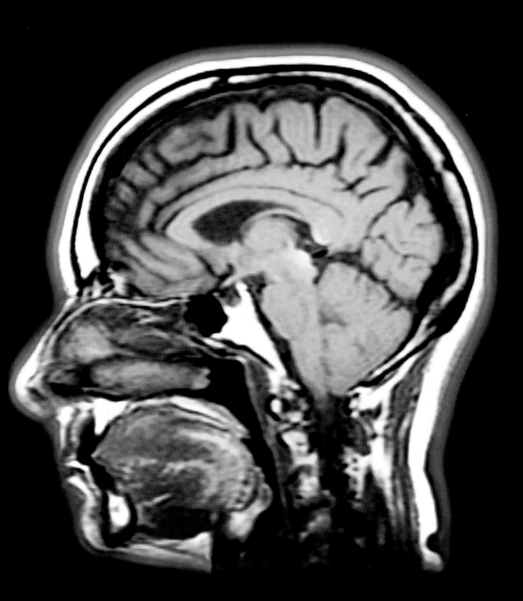

▲ The brain's cortex (outer layer) is only 5 mm thick, but flattened out would cover an area almost as big as an office desk, and contains at least 50 billion nerve cells.

The brain

- **The human brain** is made up of more than 100 billion nerve cells called neurons.

- **Each neuron** is connected to as many as 25,000 other neurons – so the brain has trillions and trillions of different pathways for nerve signals.

- **As well as controlling** our day-to-day actions and responses, the brain enables us to think, learn, understand and create.

- **The main part** of the brain is called the cerebrum and is divided into two halves. The left half controls the right side of the body, while the right half controls the left side of the body.

- **The cerebellum** at the base of the brain controls co-ordination and fine movement.

- **Girls' brains** weigh 2.5 percent of their body weight, on average, while boys' brains weigh 2 percent.

◀ Taking the top off the skull shows the brain to be a soggy, pinky-grey mass, which looks rather like a giant walnut.

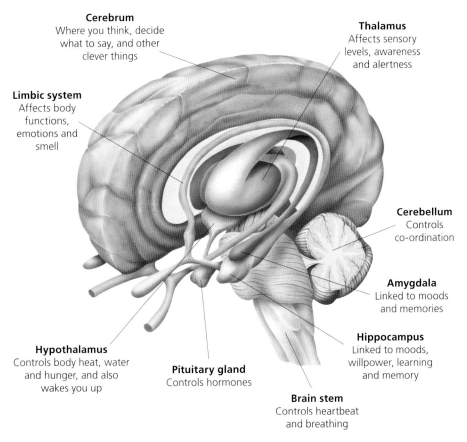

Cerebrum
Where you think, decide what to say, and other clever things

Thalamus
Affects sensory levels, awareness and alertness

Limbic system
Affects body functions, emotions and smell

Cerebellum
Controls co-ordination

Amygdala
Linked to moods and memories

Hippocampus
Linked to moods, willpower, learning and memory

Hypothalamus
Controls body heat, water and hunger, and also wakes you up

Pituitary gland
Controls hormones

Brain stem
Controls heartbeat and breathing

▲ *In this illustration, the right hemisphere (half) of the cerebrum is shown in pink, surrounding the regions that control basic drives such as hunger, thirst and anger.*

About 0.85 litres of blood shoots through your brain every minute. The brain may be as little as 2 percent of your body weight, but it demands 12–15 percent of your blood supply.

An elephant's brain weighs four times as much as the human brain. Some apes, monkeys and dolphins have a brain-body ratio quite similar to that of humans.

The cerebral cortex

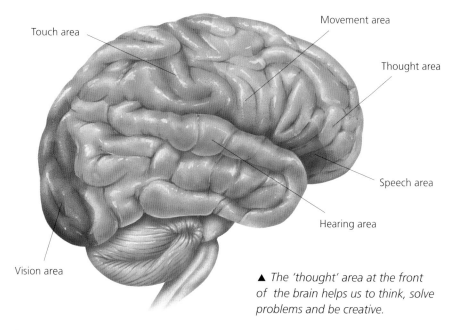

Touch area

Movement area

Thought area

Speech area

Hearing area

Vision area

▲ *The 'thought' area at the front of the brain helps us to think, solve problems and be creative.*

- **A cortex** is the outer layer of any organ, such as the brain or the kidney.

- **The cerebral cortex** is a layer of interconnected nerve cells around the outside of the brain, called 'grey matter'.

- **Conscious thoughts** and actions happen in the cerebral cortex.

- **Many signals** from the senses are registered in the brain in the cerebral cortex.

- **The visual cortex** is around the lower back of the brain. It is the place where all the things you see are registered in the brain.

- **The somatosensory cortex** is a band running over the top of the brain like a headband. This is where a touch on any part of the body is registered.

- **Just in front** of the sensory cortex lies the motor cortex. It sends out signals to body muscles to make them move.

- **The more nerve endings** there are in a particular part of the body, the more of the sensory cortex it occupies.

- **A huge proportion** of the sensory cortex is taken up by the lips and face.

- **The hands** take up almost as much of the sensory cortex as the face.

- **A human brain** has a cerebral cortex four times as big as a chimpanzee, about 20 times as big as a monkey's, and about 300 times as big as a rat's.

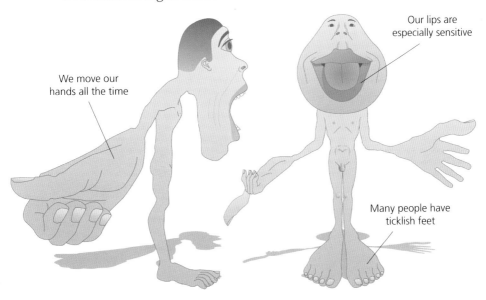

We move our hands all the time

Our lips are especially sensitive

Many people have ticklish feet

▲ These cartoons show the body in proportion to how much of the brain is needed to control movement (on the left) or process senses (on the right).

The spinal cord

- **The spinal cord** is the bundle of nerves running down the middle of the backbone.

- **It is the route** for all nerve signals travelling between the brain and the body.

- **The spinal cord** can actually work independently of the brain, sending out responses to the muscles directly.

- **The outside** of the spinal cord is made of the long tails or axons of nerve cells and is called white matter; the inside is made of the main nerve bodies and is called grey matter.

- **Your spinal cord** is about 43 cm long and one centimetre thick. It stops growing when you are about five years old.

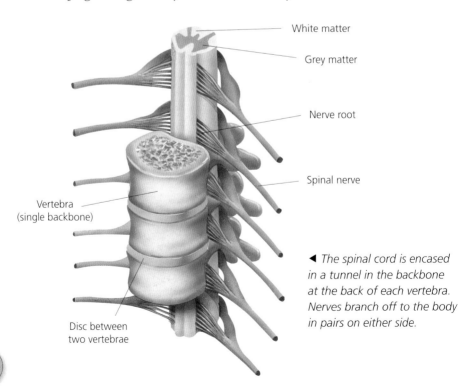

White matter

Grey matter

Nerve root

Spinal nerve

Vertebra (single backbone)

Disc between two vertebrae

◄ *The spinal cord is encased in a tunnel in the backbone at the back of each vertebra. Nerves branch off to the body in pairs on either side.*

▲ If the spinal cord is damaged, nerves cannot carry messages to and from the brain and muscles. This may mean the person cannot walk and has to use a wheelchair.

DID YOU KNOW?

The most common cause of a spinal cord injury is a traffic accident.

- **Damage to the spinal cord** can cause paralysis.

- **Injuries below the neck** can cause paraplegia – paralysis below the waist.

- **Injuries to the neck** can cause quadriplegia – paralysis below the neck.

- **Descending pathways** are groups of nerves that carry nerve signals down the spinal cord – typically signals from the brain for muscles to move.

- **Ascending pathways** are groups of nerves that carry nerve signals up the spinal cord – typically signals from the skin and internal body sensors going to the brain.

101

Peripheral nervous system

- **The peripheral nervous system** (PNS) consists of the 12 cranial nerves in the head and the 31 pairs of spinal nerves that branch off the spinal cord.

- **Spinal nerves** can be divided into groups: there are eight pairs of cervical nerves in the neck, 12 pairs of thoracic nerves in the chest, five pairs of lumbar nerves in the abdomen, five pairs of sacral nerves in the lower back and one pair of coccygeal nerves at the base of the spine.

- **Several of the spinal nerves** combine to form collections of nerves called nerve plexuses.

- **Located in the head**, the cervical plexus provides the nerves that supply the neck and shoulders.

- **The plexus** in the neck and upper arm is called the brachial plexus and supplies the arm and the upper back.

- **Nerves in the abdomen** are provided by the solar plexus. The lumbar plexus contains nerves that supply the abdomen and the leg muscles.

- **Long bundles** of nerve fibres make up the nerves of the PNS. These are in turn made from the long axons (tails) of nerve cells, bound together like the wires in a telephone cable.

- **The sciatic nerve** to each leg is the longest nerve in the body. Its name is from the Latin for 'pain in the thigh'.

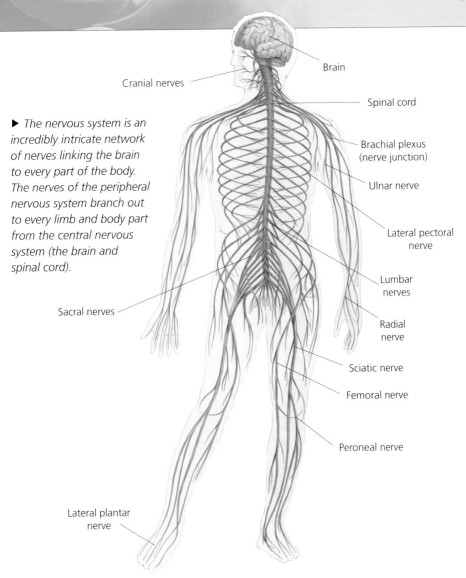

Brain

Cranial nerves

Spinal cord

▶ The nervous system is an incredibly intricate network of nerves linking the brain to every part of the body. The nerves of the peripheral nervous system branch out to every limb and body part from the central nervous system (the brain and spinal cord).

Brachial plexus (nerve junction)

Ulnar nerve

Lateral pectoral nerve

Lumbar nerves

Sacral nerves

Radial nerve

Sciatic nerve

Femoral nerve

Peroneal nerve

Lateral plantar nerve

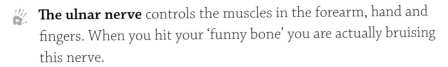

The ulnar nerve controls the muscles in the forearm, hand and fingers. When you hit your 'funny bone' you are actually bruising this nerve.

Pins and needles occur when you pinch one of your peripheral nerves by sitting awkwardly or holding an arm or leg in a funny position for a long time.

103

Cranial nerves

- **There are 12 pairs** of cranial nerves that emerge directly from the brain.

- **The nose** is linked to the brain by the olfactory nerves.

- **The eyes** are linked to the brain by the optic nerves, which carry visual information, and the oculomotor, trochlear and abducens nerves, which move the eyeballs in their sockets.

- **The trigeminal nerves** control chewing and receive sensation from the face.

- **The facial nerves** control your facial expression and carry information about your sense of taste to the brain.

- **Balance and movement** are aided by the vestibulocochlear nerves.

- **The glossopharyngeal nerves** also carry information about your sense of taste to the brain.

- **The vagus nerves** perform many functions, including controlling your heart rate and speech.

- **Some of the movements** of your neck and shoulders, such as shrugging, are controlled by the accessory nerves.

- **The last nerve**, the hypoglossal nerve, helps you to swallow and talk by controlling the tongue.

DID YOU KNOW?
Cranial means something to do with the skull. However, not all cranial nerves go to the head.

▶ *Cranial nerves come from the underside of the brain and connect the brain with the face, including the eyes, nose and mouth.*

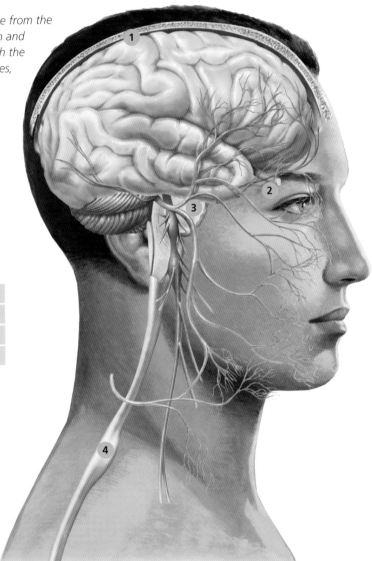

1 Skull

2 Optic nerve

3 Brain stem

4 Spinal cord

Nerve cells

- **Nerves are made** of very specialized cells called neurons.

- **Neurons are spider-shaped** with a nucleus at the centre, lots of branching threads called dendrites, and a winding tail called an axon, which can be up to one metre long.

- **Axon terminals** on the axons of one neuron link either to the dendrites or body of another neuron.

- **Neurons link up** like beads on a string to make your nervous system.

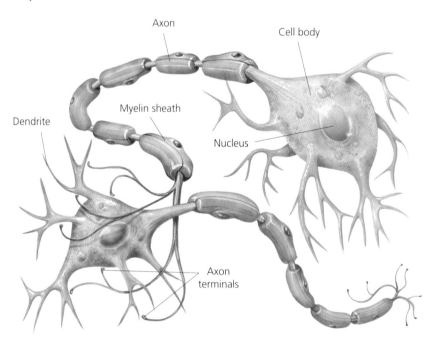

▲ Nerve cells, or neurons, are the 'wires' of the body's nervous system. They carry messages within, to and from the central nervous system along fine branches called dendrites and long tails called axons.

Most cells are short-lived and are constantly being replaced by new ones. Neurons, however, are very long-lived – some are never actually replaced once you are born.

Nerve signals travel as electrical pulses, each pulse lasting about 0.001 seconds.

When nerves are resting there are extra sodium ions with a positive electrical charge on the outside of the nerve cell, and extra negative ions inside.

When a nerve fires, little gates open in the cell wall all along the nerve, and positive ions rush in to join the negative ions. This makes an electrical pulse.

Long-distance nerves are insulated (covered) by a sheath of a fatty substance called myelin, to keep the signal strong.

Myelinated (myelin-sheathed) nerves shoot signals very fast – at more than 100 m/sec.

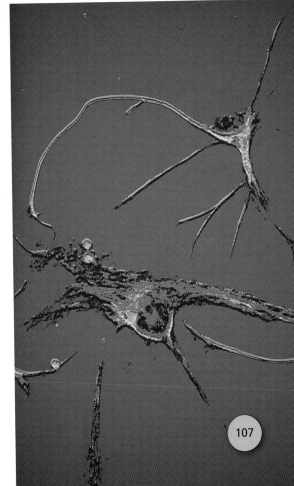

▶ *Microscopically tiny nerve cells were first seen when stained with silver nitrate by the Italian scientist Camillo Golgi in the 1870s.*

107

Synapses

- **Synapses** are the very tiny gaps between nerve cells.

- **When a nerve signal** goes from one nerve cell to another, it must be transmitted (sent) across the synapse by special chemicals called neurotransmitters.

DID YOU KNOW?

It is thought that an adult human might have as many as 500 trillion synapses in his or her brain.

- **Droplets of neurotransmitter** are released into the synapse whenever a nerve signal arrives.

- **As the droplets** of neurotransmitter lock on to the receiving nerve's receptors, they fire the signal onwards.

- **Each receptor site** on a nerve ending only reacts to certain neurotransmitters.

- **Sometimes several signals** must arrive before enough neurotransmitter is released to fire the receiving nerve.

- **More than 40** neurotransmitter chemicals have been identified.

- **Dopamine** is a neurotransmitter that works in the parts of the brain that control movement and learning. Parkinson's disease may develop when the nerves that produce dopamine break down.

- **Serotonin** is a neurotransmitter that is linked to sleeping and waking up, and also to your mood.

- **Acetylcholine** is a neurotransmitter that may be involved in memory, and also in the nerves that control muscle movement.

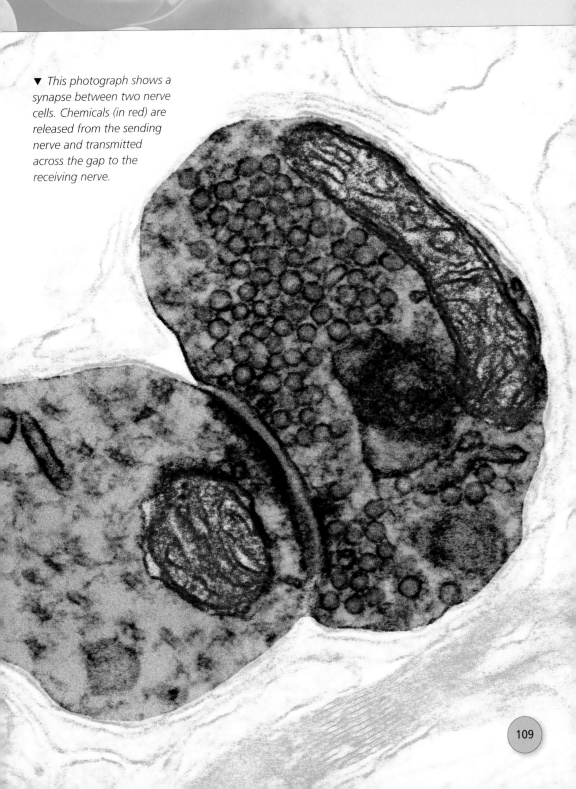

▼ *This photograph shows a synapse between two nerve cells. Chemicals (in red) are released from the sending nerve and transmitted across the gap to the receiving nerve.*

Sensory nerves

Sensory nerves are the nerves that carry information to your brain from sense receptors all over your body.

Sense receptors include the sense organs, such as the eyes, and receptors in the skin.

They are also located in the mouth, throat and in the lining of the internal organs.

▼ *We rely on sight more than our other senses, but to fully enjoy a firework display we need sight and sound.*

- **Each sense receptor** in the body is linked to the brain by a sensory nerve.

- **Most sensory nerves** feed their signals to the somatosensory cortex, which is the strip situated around the top of the brain where sensations are registered.

- **Massive bundles** of sensory nerve cells form the nerves that link major senses such as the eyes, ears and nose, to the brain.

- **Sense receptors** are everywhere in your skin, but places such as your face have more than your back.

▲ *Some of our most pleasant feelings, such as being hugged or stroked, are sent to the brain by sensory nerves.*

- **In the skin**, many sense receptors are simply 'free' – meaning they are exposed sensory nerve-endings.

- **Free nerve-endings** are rather like the bare end of a wire. They respond to all kinds of skin sensation and are almost everywhere in your skin.

- **We can tell** how strong a sensation is by how fast the sensory nerve fires signals to the brain. But no matter how strong the sensation is, the nerve does not go on firing at the same rate and soon slows down.

111

Touch

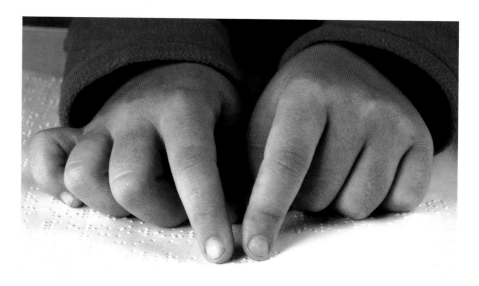

▲ *Our fingertips are most sensitive to touch, which can even compensate for lack of sight by allowing people to read Braille.*

- **Touch, or physical contact**, is just one of the five senses.

- **Touch sensors** in our skin help to relay information to the brain, enabling us to feel.

- **There are several aspects** to touch – we feel light and deep pressure, hot, cold and pain.

- **We use touch** to learn about our surroundings and carry out everyday activities such as holding things.

- **Without our** sense of touch we would not know how much pressure to use when picking something up. We would drop things or squeeze them so tight they would break.

- **Touch also** warns us of danger. We feel pain if we touch something too hot or too cold and we know to move away.

 **For blind and partially sighted people**, touch is particularly important in helping them to recognize their environment. Many blind people learn to read Braille so that they can still read books.

 **Touch is also important** for our well being. We all need to touch and be touched to grow up happy and healthy.

 **Anaesthetics** temporarily reduce or remove our sense of touch so that we don't feel pain.

 **If a nerve is damaged**, we may lose our sense of touch in the area that nerve supplies. People with diabetes sometimes lose their sense of touch in the fingertips or toes.

▶ *We can identify many things by touch alone.*

Touch sensors

- **Touch receptors** are spread all over the body in the skin. They respond to light and deep pressure, pain, hot and cold.

- **There are 200,000** hot and cold receptors in your skin, plus 500,000 touch and pressure receptors and nearly three million pain receptors.

- **There are specialized receptors** in certain places, each named after their discoverer.

- **Merkel's discs** and Meissner's corpuscles are found near the surface of the skin and react instantly to light pressure.

- **Meissner's corpuscles** are particularly numerous in sensitive hairless skin areas such as the fingers and the palms of the hands.

- **Pacinian corpuscles** lie deeper in the skin and respond to deeper pressure and vibration.

- **Ruffini's endings** also lie deeper in the skin and respond to deep and steady pressure.

- **Free nerve endings** react to pressure, pain, heat and cold.

- **When we touch something**, pressure on the skin causes the receptors to move, sending signals to the brain, which interprets the signals as light or deep pressure, pain, heat or cold.

114

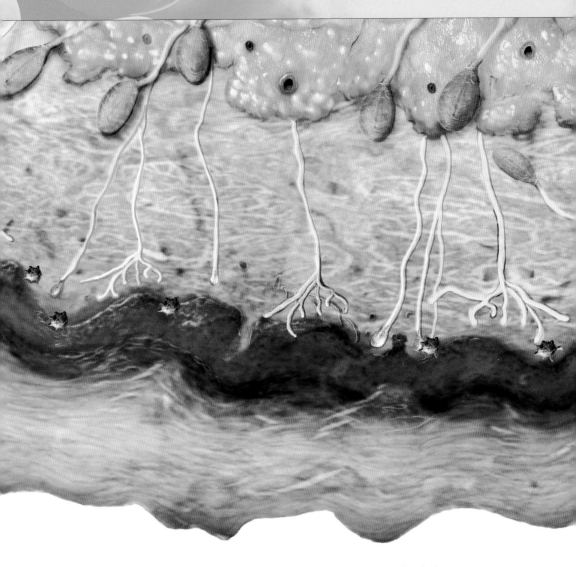

▲ *Beneath the top layer of the skin are many sensory receptors that help us to feel pain, pressure and temperature.*

It is the combination of all these receptors working together that enables us to do so many things, from hugging each other to drawing or painting.

115

The eye

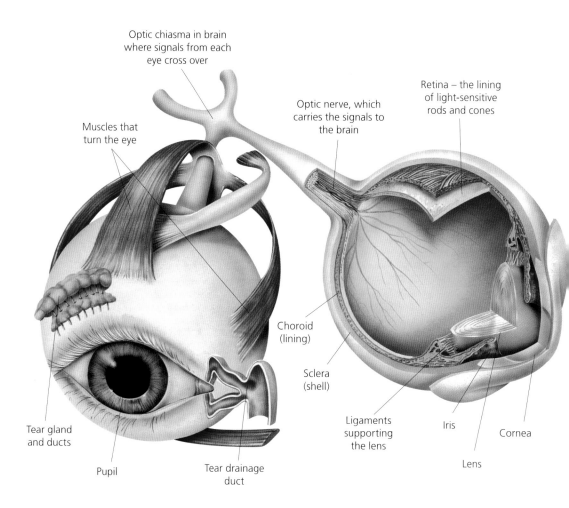

Optic chiasma in brain where signals from each eye cross over

Retina – the lining of light-sensitive rods and cones

Optic nerve, which carries the signals to the brain

Muscles that turn the eye

Choroid (lining)

Sclera (shell)

Tear gland and ducts

Ligaments supporting the lens

Iris

Cornea

Pupil

Tear drainage duct

Lens

▲ This illustration shows the two eyeballs, with a cutaway to reveal the cornea and lens (which projects light rays through the pupil) and the light-sensitive retina (which registers it).

- **Your eyes** are tough balls that are filled with a jelly-like substance called vitreous humour.

- **The eyes** are protected by bony sockets in the skull and from dust and dirt by our eyelids and eyelashes.

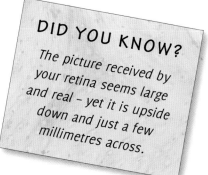

DID YOU KNOW?
The picture received by your retina seems large and real – yet it is upside down and just a few millimetres across.

- **The outer layer** of the eye, or the white of the eye, is called the sclera and helps keep the shape of the eye.

- **The cornea** is a thin, glassy dish across the front of your eye. It allows light rays through the eye's window, the pupil, and into the lens.

- **The iris** is the coloured, muscular ring around the pupil. The iris narrows in bright light and widens when light is dim.

- **The lens** is just behind the pupil. It focuses the picture of the world on to the back of the eye.

- **The back of the eye** is lined with millions of light-sensitive cells. This lining is called the retina, and it registers the picture and sends signals to the brain via the optic nerve.

- **Strong muscles** that surround each eye enable us to move our eyeballs in different directions.

- **Our eyes are kept moist** by tears, which are continuously produced by glands that lie just above each eye. Excess tears drain away from the corners of the eyes into the nose. They also help to wash away any dust from the eye.

117

Sight

- **When you look** at an object, light from the object hits the cornea at the front of your eye and is bent by the cornea on to the lens.

- **The lens** in the eye focuses the light onto the retina at the back of the eye. The most sensitive part of the retina is called the fovea.

- **The cells** in the retina collect the information and send signals to the brain, which interprets the signals as images.

- **There are two kinds** of light-sensitive cell in the retina – rods and cones. Rods are very sensitive and work in even dim light, but they cannot detect colours. Cones respond to colour.

- **Some kinds of cone** are very sensitive to red light, some to green and some to blue. One theory says that the colours we see depend on how strongly they affect each of these three kinds of cone.

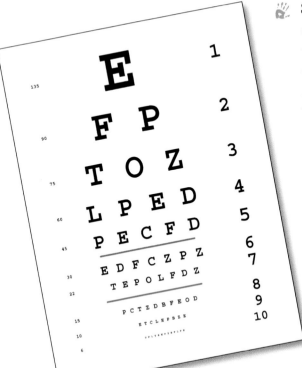

◄ An optician tests vision using a chart on a wall. The further down the chart you can read, the better your vision.

118

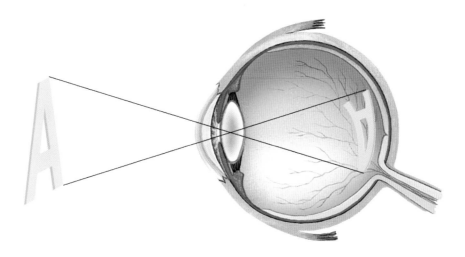

▲ *The eye produces upside-down images on the retina. Signals then travel to the brain, which interprets the image as the right way up.*

Each of your two eyes gives you a slightly different view of the world. The brain combines these views to give an impression of depth and 3D solidity.

Although each eye gives a slightly different view of the world, we see things largely as just one eye sees it. This dominant eye is usually the right eye.

Many people do not have perfect vision. People who have difficulty seeing objects that are a long way away are called short-sighted. People who have difficulty seeing objects that are very close are called long-sighted.

Short and long sight occurs when the light from the object you are looking at is not focused on the retina by the lens in the eye.

Glasses or contact lenses are used to adjust the way the light enters the eye so that the lens will focus the light on the right place in the retina.

Colour vision

Seeing in colour depends on eye cells called cones.

Cones do not work well in low light, which is why things seem grey at dusk.

Some cones are more sensitive to red light, some are more sensitive to green and some to blue.

The old trichromatic theory said that you see colours by comparing the strength of the signals from each of the three kinds of cone – red, green and blue.

The trichromatic theory does not explain colours such as gold, silver and brown.

The opponent-process theory said that you see colours in opposing pairs – blue and yellow, red and green.

In opponent-process theory, lots of blue light is thought to cut your awareness of yellow, and vice versa. Lots of green cuts your awareness of red, and vice versa.

◄ *Seeing all the colours of the world around you depends on the colour-sensitive cone cells inside your eyes.*

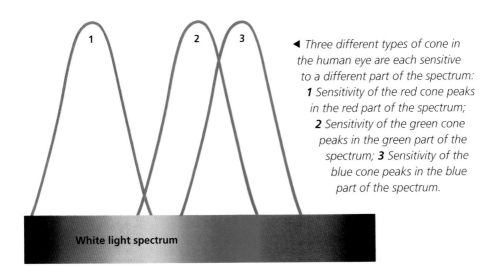

◀ Three different types of cone in the human eye are each sensitive to a different part of the spectrum: **1** Sensitivity of the red cone peaks in the red part of the spectrum; **2** Sensitivity of the green cone peaks in the green part of the spectrum; **3** Sensitivity of the blue cone peaks in the blue part of the spectrum.

White light spectrum

Now scientists combine these theories and think that colour signals from the three kinds of cone are further processed in the brain in terms of the opposing pairs.

Ultraviolet light is light in waves too short for you to see, although some birds and insects can see it.

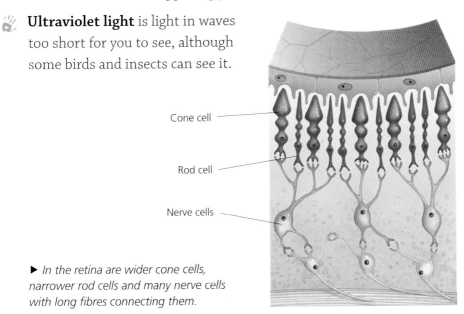

Cone cell

Rod cell

Nerve cells

▶ In the retina are wider cone cells, narrower rod cells and many nerve cells with long fibres connecting them.

121

The ear

- **Ears aren't just for hearing**; they also enable us to balance and keep our posture.

- **There are three parts** to the ear – the outer ear, which is the part you can see and the canal that leads to the eardrum, the middle ear and the inner ear.

- **Pinnae** (singular, pinna) are the ear flaps you can see on the side of your head. They are simply collecting funnels for sounds.

- **The canal** from the pinna is called the ear canal. Glands in the ear canal produce ear wax that helps to keep the ear clean.

- **The ear canal** leads to the eardrum – a thin membrane that separates the outer and middle ear.

- **Three bones** are found in the middle ear. These are the malleus (hammer), the incus (anvil) and the stapes (stirrup).

- **The stapes** is the smallest bone in the body and attaches to another membrane called the oval window, which separates the middle and inner ear.

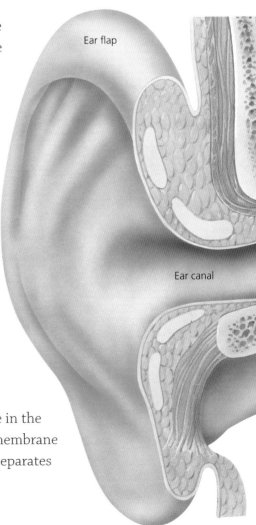

Ear flap

Ear canal

The oval window is 30 times smaller in area than the eardrum.

Beyond the oval window is the cochlea – a winding collection of three liquid-filled tubes, which looks a bit like a snail shell.

In the middle tube of the cochlea there is a flap that covers row upon row of tiny hairs. This is called the organ of Corti.

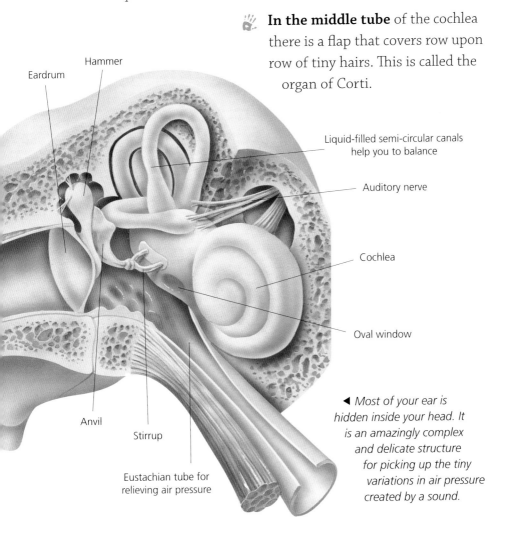

Eardrum

Hammer

Liquid-filled semi-circular canals help you to balance

Auditory nerve

Cochlea

Oval window

Anvil

Stirrup

Eustachian tube for relieving air pressure

◀ Most of your ear is hidden inside your head. It is an amazingly complex and delicate structure for picking up the tiny variations in air pressure created by a sound.

123

Hearing

🤚 **The ears** provide us with our sense of hearing. The outer part of the ear acts as a funnel for sounds, which travel down the eardrum.

🤚 **A little way** inside your head, the sounds hit a thin, tight wall of skin, called the eardrum, making it vibrate.

🤚 **When the eardrum vibrates**, it shakes the three little bones called ossicles in the middle ear.

🤚 **When the ossicles vibrate**, they rattle the tiny membrane called the oval window, intensifying the vibration.

▲ Modern hearing aids are small enough to fit inside the ear.

🤚 **The vibration** of the oval window makes waves shoot through the liquid in the cochlea in the inner ear and wash over the flap of the organ of Corti, waving it up and down.

🤚 **When the organ of Corti waves**, it tugs on the many tiny hairs under the flap. These send signals to the brain via the auditory nerve, and you hear a sound.

🤚 **There are up to** 15,000 hair cells in the Organ of Corti.

- **The average human adult** can hear frequencies between 20 and 20,000 hertz, but a bat can hear between 1000 and 120,000 hertz.

- **As we get older** we have more difficulty hearing high-pitched sounds. Hearing aids help people who cannot hear well by amplifying sounds.

- **Hearing loss** may also occur as a result of being exposed to loud noises over long periods of time, such as regularly listening to music from an MP3 player at high volumes.

▼ *Vibrations in the fluid in the inner ear are transmitted through nerves (shown in green), which run through bone (yellow) to hair cells in the organ of Corti, which send signals to the brain.*

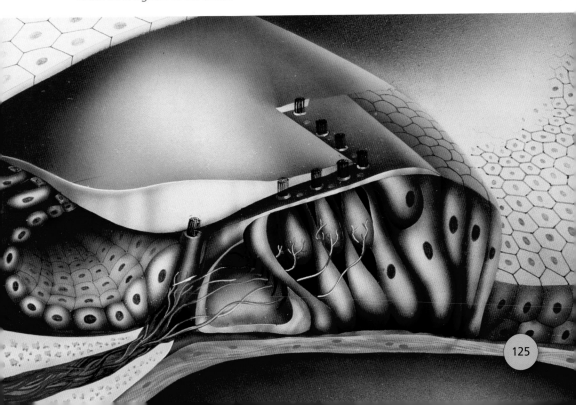

Balance

- **To stay upright**, your body must send a continual stream of data about its position to your brain – and your brain must continually tell your body how to move to keep its balance.

- **Balance is controlled** in many parts of the brain, including the brain's cerebellum.

- **Your brain** finds out about your body position from many sources, including your eyes, proprioceptors around the body, and the semicircular canals and other chambers in the inner ear.

- **Proprioceptors** are sense receptors in your skin, muscles and joints.

- **The semicircular canals** are three, tiny, fluid-filled loops in your inner ear.

- **Two chambers** (holes) called the utricle and saccule are linked to the semicircular canals.

- **When you move** your head, the fluid in the canals and cavities lags a little, pulling on hair detectors that tell your brain what is going on.

- **The canals** tell you whether you are nodding or shaking your head, and which way you are moving.

- **The utricle** and saccule tell you if you tilt your head or if its movement speeds up or slows down.

▼ A rollercoaster ride can make you feel dizzy because the liquid inside your inner ear keeps spinning after you have stopped moving.

The nose

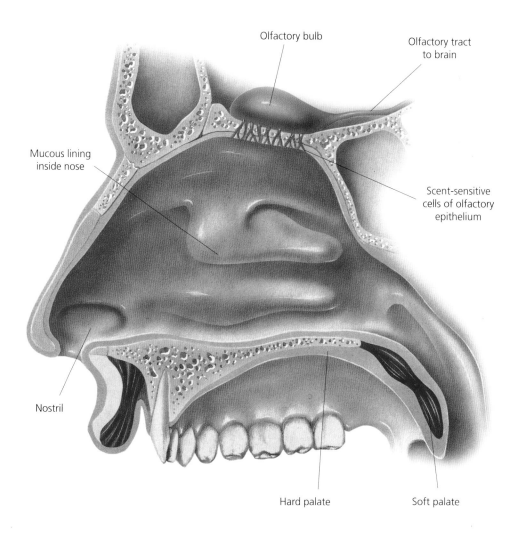

Olfactory bulb

Olfactory tract to brain

Mucous lining inside nose

Scent-sensitive cells of olfactory epithelium

Nostril

Hard palate

Soft palate

▲ *Scent particles dissolve in the mucous lining. The cells at the top of the nose then send signals along the olfactory nerve to a special part of the brain.*

- **We use our noses** to smell and to breathe in air.

- **The nose** is made of skin and cartilage attached to the ethmoid bone of the skull. The two nostrils are separated by a piece of cartilage called the nasal septum.

DID YOU KNOW?

When two people rub noses together it is called an Eskimo kiss.

- **We breathe in** through our nostrils, which open up into the nasal cavity – a large space between the mouth and the brain.

- **The nasal cavity** is lined with cells that can detect different scents and smells.

- **The human nose** can tell the difference between more than 10,000 different chemicals.

- **The nose also warms** the air that we breathe in and makes it more humid.

- **Hairs within the nose** trap dust and other particles, preventing them from entering our airways.

- **Sneezing occurs** when dust enters and irritates the lining within the nose. Sneezing may also be caused by an allergy or an infection such as a cold.

- **In some people**, sudden exposure to bright light causes sneezing.

- **In New Zealand**, pressing two noses together (called a hongi) is a traditional greeting between Maori people.

Smell

- **Smells are scent molecules** that are taken into your nose by breathed-in air. A particular smell may be noticeable even when just a single scent molecule is mixed in with millions of air molecules.

- **Dogs can pick up smells** that are 10,000 times fainter than the ones humans can detect.

- **Inside the nose**, scent molecules are picked up by a patch of scent-sensitive cells called the olfactory epithelium.

- **Olfactory** means 'to do with the sense of smell'.

- **The olfactory epithelium** contains over 25 million receptor cells.

▼ Dogs have a very keen sense of smell and are used for rescue, to sniff out drugs and for hunting.

Each of the receptor cells in the olfactory epithelium has up to 20 or so scent-detecting hairs called cilia.

When they are triggered by scent molecules, the cilia send signals to a cluster of nerves called the olfactory bulb, which then sends messages to the part of the brain that recognizes smell.

The part of the brain that deals with smell is closely linked to the parts that deal with memories and emotions. This may be why smells can often evoke vivid memories.

By the age of 20, you will have lost 20 percent of your sense of smell. By the age of 60, you will have lost 60 percent of it.

Smell is closely related to taste. If you have a cold and lose your sense of smell you will probably notice that food tastes odd or has no taste at all.

▶ *Manufacturers often use plant oils to make scented products such as soaps. Different smells are thought to create different moods. For example, the smell of lavender is thought to create a calm mood.*

Taste

 The sense of taste is the crudest of our five senses, giving us less information about the world than any other sense.

 Taste is triggered by certain chemicals in food, which dissolve in the saliva in your mouth and then send information to a particular part of the brain via sensory nerve cells on the tongue.

 Taste buds are receptor cells found around tiny bumps called papillae on the surface of your tongue.

 There are currently five distinctive recognized tastes. Umami, or savoury, as well as sweet, salty, bitter and sour.

 The back of the tongue contains big round papaillae shaped like an upside-down V, with taste buds in the deep clefts.

 The front of the tongue is where fungiform (mushroom-like) papillae and filiform (hairlike) papillae are found.

◄ *The tongue is sensitive to flavours, texture and temperature.*

► *The tongue has around 10,000 taste buds on its surface. There are also a few inside the cheeks and in the throat.*

As well as taste, the tongue can also feel the texture and temperature of food.

Your sense of taste works closely together with your sense of smell to make the flavour of food more interesting.

Strong tastes, such as spicy food, rely less on the sense of smell than on pain-sensitive nerve endings in the tongue.

People can learn to distinguish more flavours and tastes than normal, as is the case with tea- or wine-tasters.

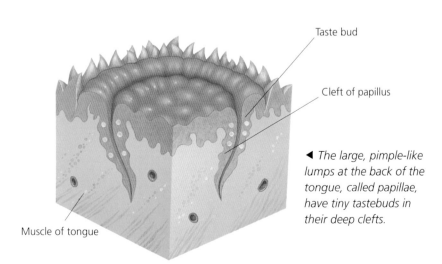

Taste bud

Cleft of papillus

◄ *The large, pimple-like lumps at the back of the tongue, called papillae, have tiny tastebuds in their deep clefts.*

Muscle of tongue

133

Motor nerves

Motor nerves are connected to your muscles and tell your muscles to move.

Each major muscle has many motor nerve endings that instruct it to contract (tighten).

Motor nerves cross over from one side of the body to the other at the top of the spinal cord. This means that signals from the right side of your brain go to the left side of your body, and vice versa.

Each motor nerve is paired to a proprioceptor on the muscle and its tendons. This sends signals to the brain to say whether the muscle is tensed or relaxed.

If the strain on a tendon increases, the proprioceptor sends a signal to the brain. The brain adjusts the motor signals to the muscle so it contracts more or less.

Motor nerve signals originate in a part of the brain called the motor cortex.

◀ When we need to act, motor nerves fire signals from the brain to the muscles to make them move.

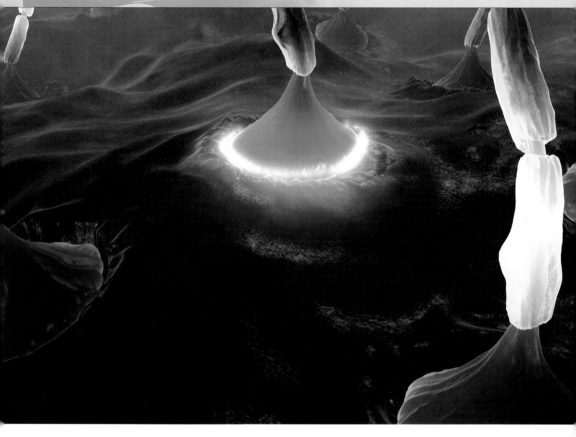

▲ *Signals from the brain (shown as a flash of green light) travel down nerve cells to a muscle, telling the muscle to contract.*

- **All the motor nerves** (apart from those in the head) branch out from the spinal cord.

- **The gut** has no motor nerve endings but plenty of sense endings, so you can feel it but cannot move it consciously.

- **The throat** has motor nerve endings but few sense endings, so you can move it but not feel it.

- **Motor neuron disease** is a disease that attacks motor nerves within the central nervous system.

Reflexes

- **Reflexes are muscle movements** that are automatic (they happen without you thinking about them).

- **Inborn reflexes** are reflexes you were born with, such as urinating, or shivering when you are cold.

- **The knee-jerk** is an inborn reflex that makes your leg jerk up when the tendon below your knee is tapped.

- **Primitive reflexes** are reflexes that babies have for a few months after they are born.

- **One primitive reflex** is when you put something in a baby's hand and it automatically grips it.

- **Reflex reactions** make you pull your hand from hot things before you have had time to think about it.

▶ *Even babies have reflexes – automatically grasping anything put into the palms of their hands.*

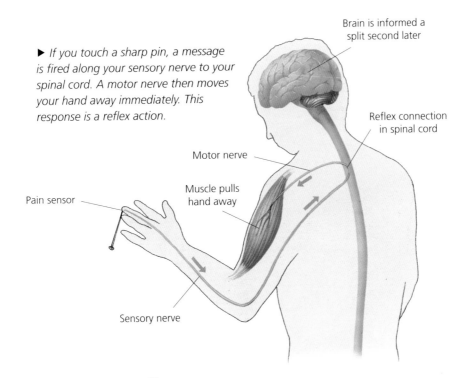

▶ *If you touch a sharp pin, a message is fired along your sensory nerve to your spinal cord. A motor nerve then moves your hand away immediately. This response is a reflex action.*

Brain is informed a split second later

Reflex connection in spinal cord

Motor nerve

Muscle pulls hand away

Pain sensor

Sensory nerve

DID YOU KNOW?

Athletes often have lightning reflexes – their bodies react faster than their brains can process.

Reflex reactions work by short-circuiting the brain. The alarm signal from your hand sets off motor signals in the spinal cord to move the hand.

A reflex arc is the nerve circuit from sense to muscle via the spinal cord.

Conditioned reflexes are those you learn through habit, as certain pathways in the nervous system are used again and again.

They can help you do anything, from holding a cup to playing football, without thinking.

Co-ordination

◀ Dance of any kind tests the body's balance and co-ordination to its limits.

- **Co-ordination** means balanced or skilful movement.

- **To make you move**, your brain has to send signals out along nerves telling all the muscles involved exactly what to do.

- **Co-ordination of the muscles** is handled by the cerebellum at the back of your brain.

▲ *Any sport requires high degrees of muscle co-ordination. Team sports also need you to co-ordinate your movements exactly with other team members.*

- **The cerebellum** is given instructions by the brain's motor cortex.

- **The cerebellum's commands** are sent via the basal ganglia in the middle of the brain.

- **Proprioceptors are nerve cells** that are sensitive to movement, pressure or stretching. Proprioceptor means 'one's own sensors'.

- **They are all over your body** – in muscles, tendons and joints – and they all send signals to your brain telling it the position or posture of every part of your body.

- **The hair cells** in the balance organs of your ear are also proprioceptors.

Thinking

Some scientists claim that we humans are the only living things that are conscious, meaning that we alone are actually aware that we are thinking.

No one knows how consciousness works – it is one of science's last great mysteries.

Most of your thoughts seem to take place in the cerebrum (at the top of your brain), and different kinds of thought are linked to different areas, called association areas.

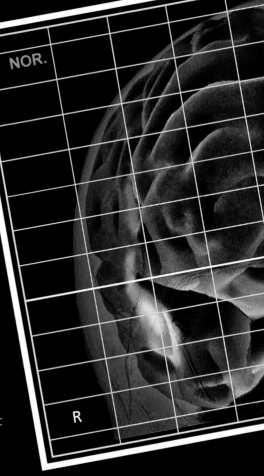

NOR.

R

Each half of the cerebrum has four rounded ends called lobes – two at the front, called frontal and temporal lobes, and two at the back, called occipital and parietal lobes.

The frontal lobe is linked to your personality and it is where you have your bright ideas.

The temporal lobe is where you hear and understand what people are saying to you.

▼ *This scan shows brain activity in someone reading aloud. The coloured active area on the right is the part of the brain that processes language.*

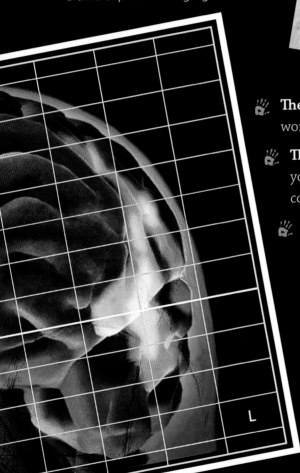

The occipital lobe is where you work out what your eyes see.

The parietal lobe is where you register touch, heat and cold, and pain.

The left half of the brain (left hemisphere) controls the right side of the body. The right half (right hemisphere) controls the left side.

One half of the brain is always dominant (in charge). Usually, the left brain is dominant, which is why up to 90 percent of people are right-handed.

141

Memory

- **When you remember something**, your brain probably stores it by creating new nerve connections.

- **You have three types** of memory – sensory, short-term and long-term.

- **Sensory memory** is when you go on feeling a sensation for a moment after it stops.

- **Short-term memory** is when the brain stores things for a few seconds, like a phone number you remember long enough to press the buttons.

- **Long-term memory** is memory that can last for months or maybe even your whole life.

- **Your brain** seems to have two ways of remembering things for the long term. Scientists call these two different ways declarative and non-declarative memories.

- **Non-declarative memories** are skills you teach yourself by practising, such as playing badminton or the flute. Repetition establishes nerve pathways.

- **Declarative memories** are either episodic or semantic. Each may be sent by the hippocampus region of the brain to the correct place in the cortex, the brain's wrinkly outer layer, where you do most of your thinking.

HOW TO IMPROVE YOUR MEMORY

CHUNKING	To remember a long string of numbers, break it down into smaller chunks
SPACING IT OUT	Don't cram – space out your learning rather than trying to remember everything at once
CUES	Use something you do every day to help you remember other things
IMAGERY	Visualising images as you are learning can help you to remember items or names
SELF-REFERENCE	It is easier to remember something if you can link it with something personal
PLACE THINGS	To remember a collection of things, try imagining them in specific places in your home

- **Episodic memories** are memories of striking events in your life, such as breaking your leg or your first day at a new school. You not only recall facts, but sensations.

- **Semantic memories** are facts such as dates. The brain seems to store these in the left temporal lobe, at the front left-hand side of the brain.

Mood

- **Mood is your state of mind** – whether you are happy or sad, angry or afraid, overjoyed or depressed.

- **Moods and emotions** seem to be strongly linked to the structures in the centre of the brain, where unconscious activities are controlled.

- **Moods have three elements** – how you feel, what happens to your body, and what moods make you do.

- **Some scientists think** the way you feel causes changes in the body – you are happy so you smile, for example.

- **Other scientists think** changes in the body alter the way you feel – smiling makes you happy.

▲ *Scientists are only just beginning to discover how moods and emotions are linked to particular parts of the brain.*

- **Yet other scientists** think moods start automatically – before you even know it – when something triggers off a reaction in the thalamus in the centre of the brain.

- **The thalamus** then sends mood signals to the brain's cortex and you become aware of the mood.

- **It also sets off** automatic changes in the body through the nerves and hormones.

- **Certain memories** or experiences are so strongly linked in your mind that they can often automatically trigger a certain mood.

DID YOU KNOW?
Every mood is a combination of how much energy you have and how stressed you are.

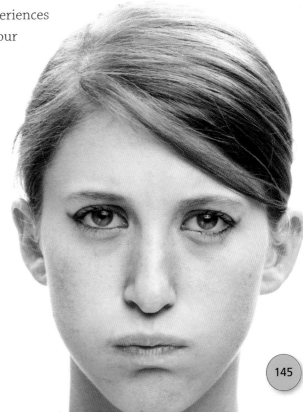

▶ Why we react the way we do is unclear, but it is emotions such as happiness or frustration that make us unique as human beings.

145

Fear and phobias

▶ *Some people have an intense fear or phobia of spiders.*

- **Fear is an emotion** that occurs in response to danger. It can vary from mild fear, such as worry over an exam, to panic.

- **Fear is a normal reaction** to a difficult situation and may help us avoid danger or cope with a threat.

- **A phobia** is an intense fear of a particular situation or thing. Phobias are extreme reactions and often irrational.

- **Someone with a phobia** will have a persistent fear of something that other people will not be afraid of, such as water or flying in an aeroplane.

- **Some phobias** may develop after a traumatic experience. For example, someone who is bitten by a dog may then develop a phobia of dogs.

- **Between 10 and 20 percent** of people have a phobia. Most phobias do not interfere with everyday lives.

- **Social phobia** is the fear of meeting new people and of social situations.

- **Claustrophobia** is the fear of small spaces; agoraphobia is the fear of leaving home or of wide, open, unfamiliar spaces.

- **A fear of spiders** is called arachnophobia. The fear of snakes is called ophidiophobia.

- **Trypanophobia** is the fear of needles and injections. People who faint at the sight of blood may have this type of phobia.

▼ *Fear can be exciting – bungee jumping gives us an adrenaline rush because it is dangerous.*

147

Communication

- **We communicate** in many different ways, not just through speech but also through our facial expressions, tone of voice and our body language.

- **Although we are the only animals** to use a complicated spoken language, all animals use sounds and body language to communicate.

- **We use our vocal cords** to speak, and we interpret other people's speech using our brains. However, it is not just what we say that is important. Our brains also interpret the tones of people's voices to help us understand what is meant.

- **Your expression,** posture and pose all contribute to your body language. It is thought that up to 70 percent of human communication is through body language.

- **Body language** can reveal whether you are telling the truth. Poker players use body language to tell if a person is bluffing or not.

▼ *A crowd at a sports event reveal their excitement and anticipation through their body language.*

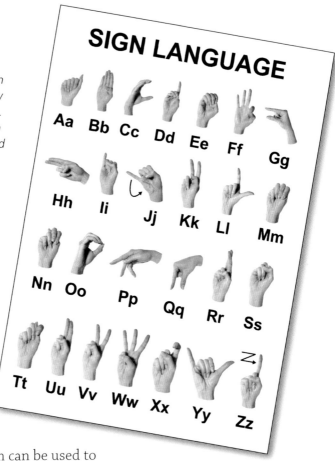

▶ *In its simplest form, sign language can spell out any word just using the hands. Hundreds of different sign languages are used around the world.*

Body language can also reveal whether you are bored, confident, anxious or interested in someone.

It is often easy to tell someone's mood just by looking at their face. We have over 30 muscles in our face – all of which can be used to make a huge variety of facial expressions.

Sign language is an important form of communication for people who cannot hear. A combination of hand, body and facial gestures is used to express words and phrases.

Sign language is not just for deaf people – divers use sign language to communicate with each other under water and soldiers use sign language when it is important to be quiet.

149

Sleeping

When you are asleep, many of your body functions go on as normal – even your brain goes on receiving sense signals. But your body may save energy and do routine repairs.

Lack of sleep can be dangerous. A newborn baby needs 18–20 hours of sleep a day. An adult needs around seven to eight hours.

Sleep is controlled in the brain stem. Dreaming is stimulated by signals fired from a part of the brain stem called the pons.

When you are awake, there is little pattern to the electricity created by the firing of the brain's nerve cells. But as you sleep, more regular waves appear.

While you are asleep, alpha waves sweep across the brain every 0.1 seconds. Theta waves are slower.

For the first 90 minutes of sleep, your sleep gets deeper and the brain waves become stronger.

After about 90 minutes of sleep, your brain suddenly starts to buzz with activity, yet you are hard to wake up.

- **After 90 minutes of sleep,** your eyes begin to flicker from side to side under their lids. This is called Rapid Eye Movement (REM) sleep.

- **REM sleep** is thought to show that you are dreaming.

- **While you sleep,** ordinary deeper sleep alternates with spells of REM lasting up to half an hour.

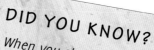

DID YOU KNOW?

When you sleep is partly controlled by your body's internal clock – the circadian clock.

▼ *As we grow, we need less sleep. Toddlers and young children need around 11–15 hours a day.*

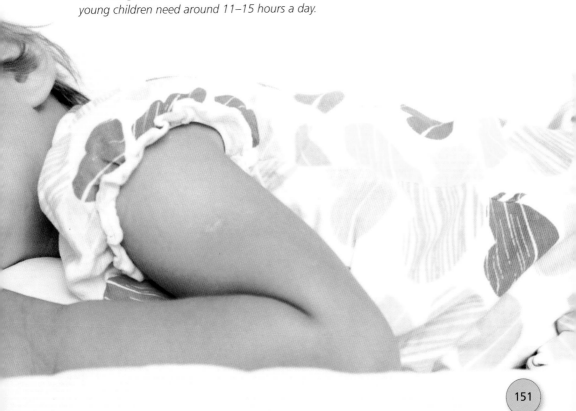

Circulatory system

Circulation

- **Your circulation** is the system of tubes called blood vessels that carries blood out from your heart to all your body cells and back again.

- **Blood circulation** was discovered in 1628 by the English physician William Harvey (1578–1657), who built on the ideas of the Italian anatomist Matteo Colombo (*c.* 1516–1559).

- **Each of the body's** 600 billion cells gets fresh blood continuously, although the blood flow is pulsating.

- **On the way out** from the heart, blood is pumped through vessels called arteries and arterioles.

- **On the way back** to the heart, blood flows through venules and veins. For each outward-going artery there is usually an equivalent returning vein.

- **Blood flows** from the arterioles to the venules through the tiniest tubes called capillaries.

- **The blood circulation** has two parts – the pulmonary and the systemic.

- **The pulmonary circulation** is the short section that carries blood low in oxygen from the right side of the heart to the lungs for 'refuelling'. It then returns oxygen-rich blood to the left side of the heart.

- **The systemic circulation** carries oxygen-rich blood from the left side of the heart all around the body. It returns blood that is low in oxygen to the right side of the heart.

- **Inside the blood**, oxygen is carried by the haemoglobin in red blood cells.

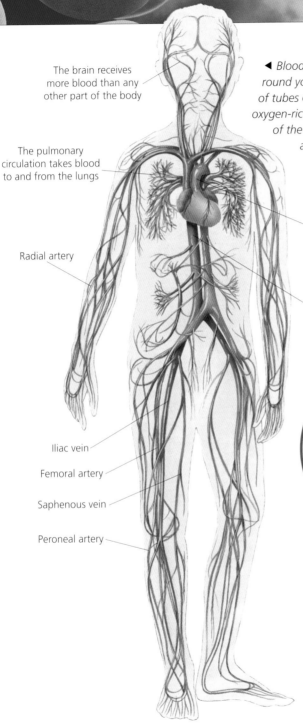

The brain receives more blood than any other part of the body

The pulmonary circulation takes blood to and from the lungs

Radial artery

Iliac vein

Femoral artery

Saphenous vein

Peroneal artery

◀ *Blood circulates continuously round and round your body, through an intricate series of tubes called blood vessels. Bright red, oxygen-rich blood is pumped from the left side of the heart through vessels called arteries and arterioles. Purplish-blue, low-in-oxygen blood returns to the right of the heart through veins and venules.*

Blood leaves the left side of the heart through a giant artery called the aorta

Blood returns to the heart through main veins called the vena cavae

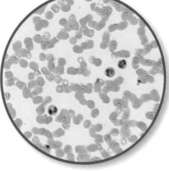

▲ *Red blood cells can actually be brown in colour, but they turn bright scarlet when their haemoglobin is carrying oxygen. After the haemoglobin passes its oxygen to a cell, it fades to dull purple.*

The heart

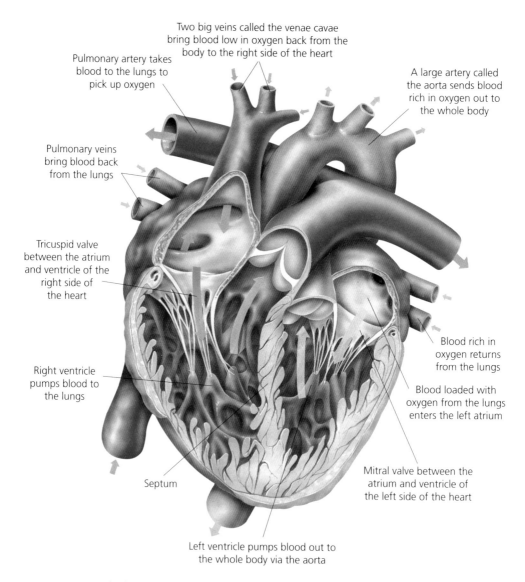

Two big veins called the venae cavae bring blood low in oxygen back from the body to the right side of the heart

Pulmonary artery takes blood to the lungs to pick up oxygen

A large artery called the aorta sends blood rich in oxygen out to the whole body

Pulmonary veins bring blood back from the lungs

Tricuspid valve between the atrium and ventricle of the right side of the heart

Blood rich in oxygen returns from the lungs

Blood loaded with oxygen from the lungs enters the left atrium

Right ventricle pumps blood to the lungs

Mitral valve between the atrium and ventricle of the left side of the heart

Septum

Left ventricle pumps blood out to the whole body via the aorta

▲ *The heart is a remarkable double pump, with two pumping chambers, the left and the right ventricles. It contracts automatically to squeeze jets of blood out of the ventricles and through the arteries.*

DID YOU KNOW?

During an average lifetime the heart will pump 200 million litres of blood.

- **Your heart** is the size of your fist. It is inside the middle of your chest, slightly to the left.

- **The heart** is a powerful pump made almost entirely of muscle.

- **The heart contracts** (tightens) and relaxes automatically about 70 times a minute to pump blood out through your arteries.

- **The heart** has two sides separated by a muscle wall called the septum.

- **The right side** is smaller and weaker, and it pumps blood only to the lungs.

- **The stronger left side** pumps blood around the body.

- **Each side of the heart** has two chambers. There is an atrium (plural atria) at the top where blood accumulates (builds up) from the veins, and a ventricle below that contracts to pump blood out into the arteries.

- **The ventricles** have much thicker muscular walls and are much stronger than the atria as they have to pump blood further around the body.

- **Each side of the heart** (left and right) ejects about 70 ml of blood every beat.

- **The coronary arteries** supply the heart. If they become clogged, the heart muscle may be short of blood and stop working. This is what happens in a heart attack.

157

Heart valves

- **There are two valves** in each side of the heart to make sure that blood flows only one way.

- **The mitral valve** between the left atrium and left ventricle is also known as the bicuspid valve as it has two flaps.

- **The valve** between the right atrium and the right ventricle is called the tricuspid valve as it has three flaps.

- **When the ventricles contract**, the blood in the chamber pushes back against the flaps of the valves, closing them and preventing blood from flowing back into the atria.

- **The aortic valve** guards the exit from the left ventricle into the aorta.

- **The pulmonary valve** guards the exit from the right ventricle, which leads into the pulmonary artery.

- **Known together** as semilunar valves, the aortic and pulmonary valves prevent blood flowing back into the two ventricles.

- **The closing** of the heart valves creates the 'lub-dup' sound of the heartbeat. The mitral and tricuspid valves close first to make the 'lub' sound; the semilunar valves close next to make the 'dup' sound.

- **Heart valves** are put under a lot of stress and strain and may become damaged or diseased. Occasionally a valve may not form properly and not work correctly from birth.

DID YOU KNOW?

When a doctor listens to your heart it is actually the sound of the heart valves closing that he or she hears.

Damaged valves can be replaced. Replacement valves may come from donors or pigs or a mechanical heart valve may be implanted.

▼ *This coloured scan shows an artificial heart valve (white, at the centre) where the main artery in the body, the aorta, enters the heart. The stitches used to close the chest can be seen at upper left.*

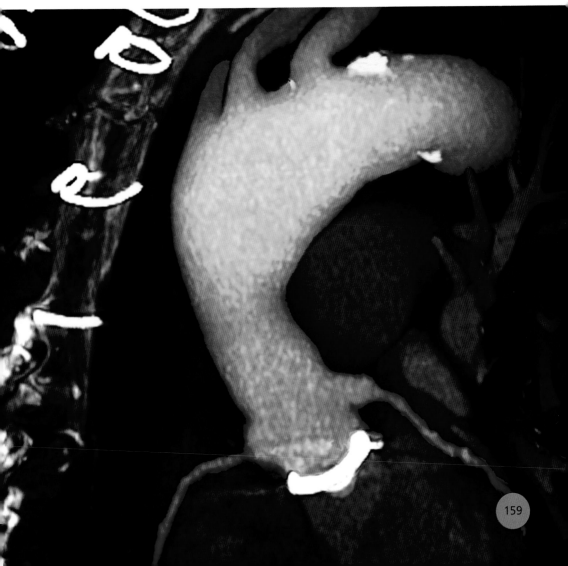

Heartbeat

- **The heartbeat** is the regular squeezing of the heart muscle to pump blood around the body.

- **It is also the term** given to the 'lub-dup' sound that the heart makes when it beats. This sound can be heard through a stethoscope.

- **The heartbeat sequence** is called the cardiac cycle and it has two phases – systole and diastole.

- **Systole is when** the heart muscle contracts (tightens). Diastole is the resting phase between contractions.

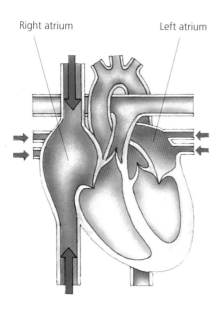

Right atrium   Left atrium

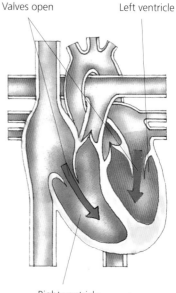

Valves open   Left ventricle

Right ventricle

1 *Blood floods into the relaxed atria.*

2 *The wave of contraction squeezes blood into the ventricles.*

- **Systole begins** when a wave of muscle contraction sweeps across the heart and squeezes blood from each of the atria into the two ventricles.

- **When the contraction** reaches the ventricles, they squeeze blood out into the arteries.

- **In diastole**, the heart muscle relaxes and the atria fill with blood again.

- **Heart muscle** on its own would contract automatically.

- **Nerve signals** make the heart beat faster or slower.

Blue: deoxygenated blood to lungs

Red: oxygenated blood to body

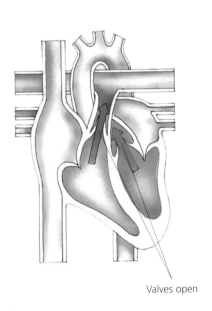

Valves open

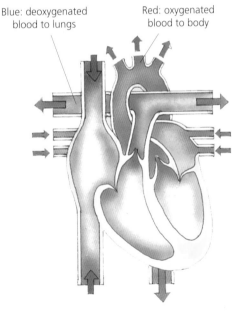

3 *Blood is squeezed out of the ventricles into the arteries.*

4 *Blood starts to fill up the now relaxed atria again.*

161

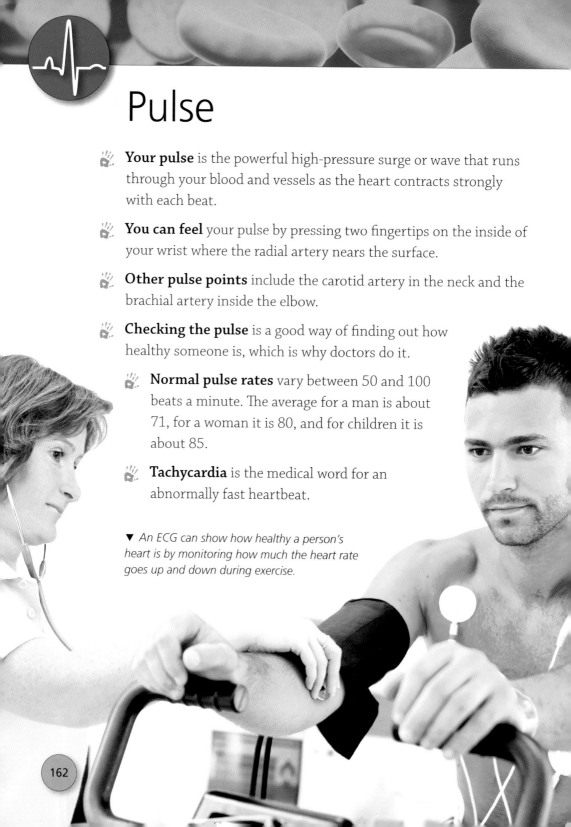

Pulse

- **Your pulse** is the powerful high-pressure surge or wave that runs through your blood and vessels as the heart contracts strongly with each beat.

- **You can feel** your pulse by pressing two fingertips on the inside of your wrist where the radial artery nears the surface.

- **Other pulse points** include the carotid artery in the neck and the brachial artery inside the elbow.

- **Checking the pulse** is a good way of finding out how healthy someone is, which is why doctors do it.

- **Normal pulse rates** vary between 50 and 100 beats a minute. The average for a man is about 71, for a woman it is 80, and for children it is about 85.

- **Tachycardia** is the medical word for an abnormally fast heartbeat.

▼ An ECG can show how healthy a person's heart is by monitoring how much the heart rate goes up and down during exercise.

▲ *A doctor measures a patient's pulse by timing how many beats there are in a minute.*

- **Someone who has tachycardia** when sitting down may have drunk too much coffee or tea, or taken drugs, or be suffering from anxiety or a fever, or have heart disease.

DID YOU KNOW?

A fit athlete usually has a pulse rate of only 40–60 beats a minute.

- **Bradycardia** is an abnormally slow heartbeat rate.

- **Arrhythmia** is any abnormality in a person's heartbeat rate.

- **Anyone with a heart problem** may be connected to a machine called an electrocardiogram (ECG) to monitor (observe) their heartbeat.

163

Heart disease

- **Your heart beats** on average 70 times a minute and an average of 2.5 billion times in a lifetime.

- **Heart disease** is the main cause of death in many Western countries, including the UK and USA.

- **For the heart** to keep beating steadily, it needs a good supply of oxygen from its own blood supply. The arteries that supply the heart are called coronary arteries.

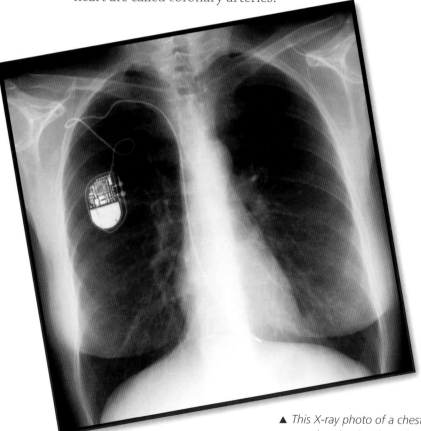

▲ *This X-ray photo of a chest shows a pacemaker that has been implanted to control an irregular heartbeat.*

- **Fatty deposits** in the coronary arteries can block the blood supply to the heart, causing chest pain or a heart attack.

- **In some cases**, surgery to widen the arteries will be successful in restoring the blood supply to the heart.

- **In other cases** surgery will be needed to provide the heart with new coronary arteries.

- **The rate** at which the heart beats is controlled by a small area in the heart, which sends out electrical impulses. If this area becomes damaged, the heart may start to beat irregularly, or even occasionally stop beating.

▲ Eating fatty foods, such as cheeseburgers, can cause fat to build up in blood vessels, which may lead to a heart attack.

- **Pacemakers** can be implanted to correct the heartbeat so that it is regular again.

- **Occasionally the heart muscle** may not work effectively. These people may need a heart transplant.

165

Arteries

- **An artery** is a tube-like blood vessel that carries blood away from the heart.

- **Systemic arteries** deliver oxygenated blood around the body. Pulmonary arteries deliver deoxygenated blood to the lungs.

- **An arteriole** is a smaller branch off an artery. Arterioles branch into microscopic capillaries.

- **Blood flows** through arteries at 30 cm per second in the main artery, down to 2 cm or less per second in the arterioles.

- **Arteries run alongside** most of the veins that return blood to the heart.

- **The walls of arteries** are muscular and can expand or relax to control the blood flow.

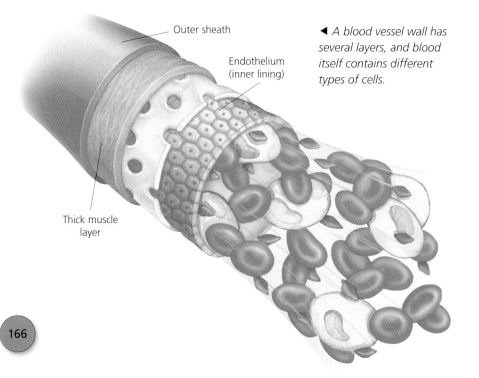

Outer sheath

Endothelium (inner lining)

Thick muscle layer

◄ A blood vessel wall has several layers, and blood itself contains different types of cells.

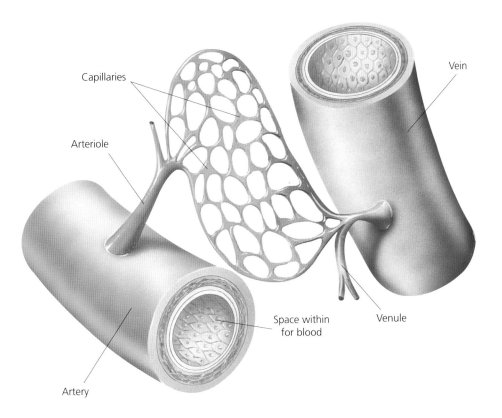

Capillaries

Arteriole

Vein

Space within
for blood

Venule

Artery

▲ *The two main kinds of blood vessel are arteries (red) and veins (blue). An artery branches into tiny capillaries, which join up to supply the vein.*

- **Arteries have thicker**, stronger walls than veins, and the pressure of the blood in them is a lot higher.

- **Over-thickening** of the artery walls may be one of the causes of hypertension (high blood pressure).

- **In old age** the artery walls can become very stiff. This hardening of the arteries (arteriosclerosis) can cut blood flow to the brain.

Capillaries

- **Capillaries** are the smallest of all your blood vessels, only visible under a microscope. They link the arterioles to the venules.

- **Capillaries were discovered** by Marcello Malphigi in 1661.

- **There are ten billion** capillaries in your body.

- **The largest capillary** is just 0.2 mm wide – thinner than a hair.

- **Each capillary** is about 0.5 mm to 1 mm long.

- **Capillary walls** are just one cell thick, so it is easy for chemicals to pass through them.

▶ You generate heat when exercising, which the body tries to lose by opening up capillaries in the skin, turning it red.

It is through the capillary walls that your blood passes oxygen, food and waste to and from each one of your body cells.

There are many more capillaries in active tissues such as muscles, liver and kidneys than there are in tendons and ligaments.

Capillaries carry less or more blood according to need. They carry more to let more blood reach the surface when you are warm. They carry less to keep blood away from the surface and save heat when you are cold.

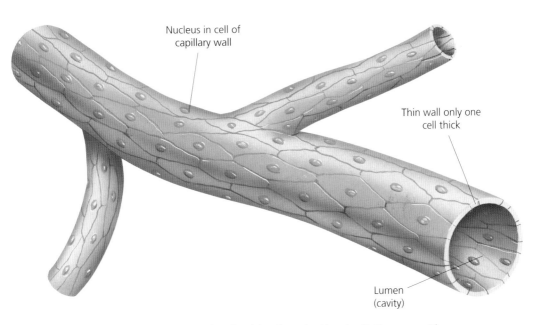

Nucleus in cell of capillary wall

Thin wall only one cell thick

Lumen (cavity)

▲ Capillaries are tiny tubes, barely wider than the blood cells they carry. They form an extensive network that twists and turns through the body's tissues.

169

Veins

- **Veins are pipes** in the body for carrying blood back to the heart.

- **Unlike arteries**, most veins carry 'used' blood back to the heart – the body cells have taken the oxygen they need from the blood, so it is low in oxygen.

- **When blood is low** in oxygen, it is a dark, purplish blue colour – unlike the bright red of the oxygenated blood carried by the arteries.

- **The only veins** that carry oxygenated blood are the four pulmonary veins, which carry blood from the lungs the short distance to the heart.

- **The two largest** veins in the body are the vena cavae, which flow into the heart from above and below.

KEY VEINS
1 Jugular vein
2 Brachial vein
3 Pulmonary vein
4 Vena cava

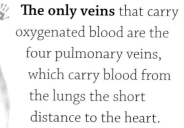

◄ Veins carry blood from all around the body back to the heart.

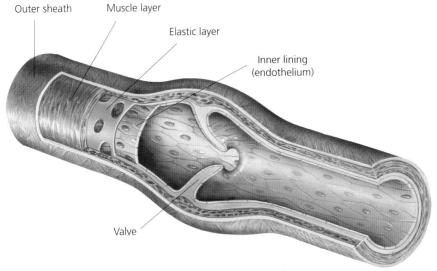

Outer sheath
Muscle layer
Elastic layer
Inner lining
(endothelium)
Valve

▲ *This shows a greatly enlarged cutaway of a small vein. The valve prevents the blood from flowing backwards away from the heart.*

- **Inside most veins** are flaps that act as valves to make sure that the blood only flows one way.

- **The blood in veins** is pumped by the heart, but the blood pressure is much lower than in arteries, and vein walls do not need to be as strong as those of arteries.

- **Unlike arteries**, veins collapse when empty.

- **Blood is helped** through the veins by pressure that is placed on the vein walls by the surrounding muscles.

> **DID YOU KNOW?**
> At any moment, 75 percent of the body's blood is in the veins.

171

Blood

- **Blood is the reddish liquid** that circulates around your body. It carries oxygen and food to body cells, and takes carbon dioxide and other waste away. It fights infection, keeps you warm, and distributes chemicals that control body processes.

- **Blood is made up** of red cells, white cells and platelets, all carried in a liquid called plasma.

DID YOU KNOW?

Oxygen in the air turns blood bright red when you bleed. In your veins it can be almost brown.

▼ *A centrifuge is used to separate the different components of blood. The spinning action of the machine separates the heavier blood cells from the lighter plasma.*

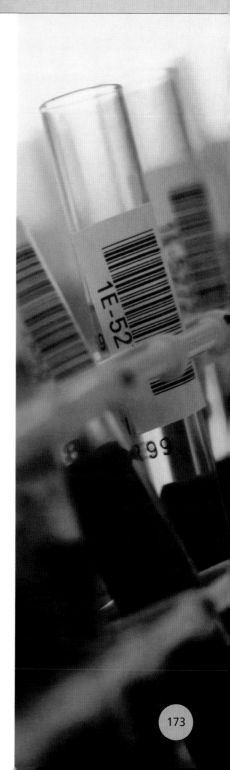

▶ *Blood contains millions of cells, carried in a clear, straw-coloured liquid called plasma.*

- **Plasma is 90 percent water**, plus hundreds of other substances, including nutrients, hormones and special proteins for fighting infection.

- **Blood plasma** turns milky immediately after a meal high in fats.

- **Platelets are tiny** pieces of cell that make blood clots start to form in order to stop bleeding.

- **The amount of blood** in your body depends on your size. An adult who weighs 80 kg has about 5 litres of blood. A child who is half as heavy has half as much blood.

- **A drop of blood** the same size as the dot on this 'i' contains around five million red cells.

- **If a blood donor** gives 0.5 litres of blood, the body replaces the plasma in a few hours, but it takes a few weeks to replace the red cells.

- **It takes about one minute** for your blood to circulate around your body. If you are exercising and your heart is beating fast, blood can circulate right round your body in about 20 seconds.

Blood cells

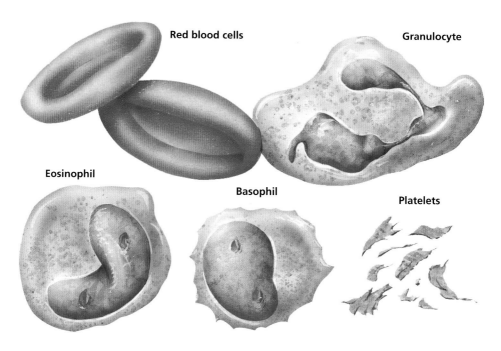

Red blood cells

Granulocyte

Eosinophil

Basophil

Platelets

▲ *These are some important kinds of cell in the blood – red cells, three kinds of white cells, and platelets.*

Your blood has two main kinds of cell – red cells and white cells – plus pieces of cell called platelets.

Red cells are button-shaped and they contain mainly a red protein called haemoglobin.

Haemoglobin is what allows red blood cells to ferry oxygen around your body.

Red cells also contain enzymes, which the body uses to make certain chemical processes happen.

- **White blood cells** are big cells called leucocytes and most types are involved in fighting infections.

- **Most white cells** contain tiny little grains and are called granulocytes.

- **Most granulocytes** are giant white cells called neutrophils. They are the blood's cleaners, and their task is to eat up invaders.

- **Eosinophils and basophils** are granulocytes that are involved in fighting disease. Some release antibodies that help fight infection.

- **Lymphocytes** are also types of white cells.

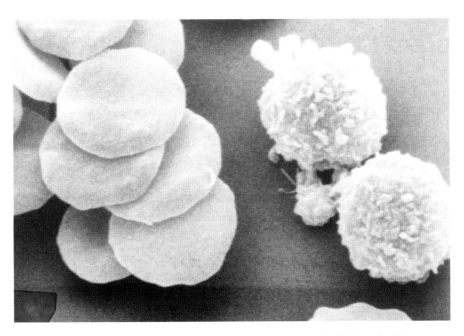

▲ *This is a highly magnified photograph of red blood cells (left) and white blood cells.*

Blood groups

Most people's blood belongs to one of four different groups or types – A, O, B and AB.

Blood type O is the most common, followed by blood group A.

Blood is also either Rhesus positive (Rh+) or Rhesus negative (Rh-).

Around 85 percent of people are Rh+. The remaining 15 percent are Rh-.

If your blood is Rh+ and your group is A, your blood group is said to be A positive. If your blood is Rh- and your group is O, you are O negative, and so on.

The Rhesus factors got their name because they were first identified in Rhesus monkeys. (The Rhesus macaque is a species of monkey used extensively in medical research.)

◄ *Blood donors usually donate about 500 ml of blood at a time and are able to give blood about every two months.*

▲ Donated blood is always tested to determine its blood group. Patients who have a blood transfusion must receive blood that matches their own group.

A transfusion is when you are given blood from another person's body. Your blood is 'matched' with other blood considered safe for transfusion.

Blood transfusions are given when someone has lost too much blood due to an injury or operation. They are also given to replace diseased blood.

DID YOU KNOW?

People with 0 negative blood are called universal donors because their blood can be given to anyone else without causing a reaction.

177

Wound healing

- **When we injure ourselves**, the damage causes a series of chemical reactions.

- **White blood cells** appear at the scene of the wound to fight infection.

- **Platelets** start to stick to each other and to the walls of the damaged blood vessels.

- **The sticky platelets** attract more platelets, forming a plug to stop you losing blood.

- **The platelets release** a sequence of chemicals called clotting factors (factors 1 through to 8).

- **At the final stage** of the clotting sequence a lacy, fibrous network is formed from a protein called fibrin.

- **The fibrin traps** red blood cells to form a blood clot that seals the damaged vessel.

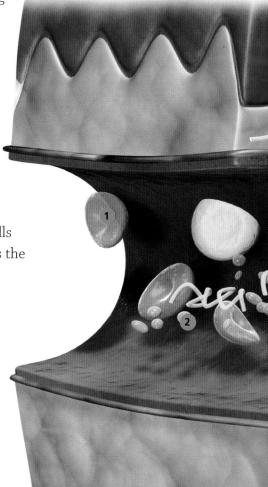

▶ When you are injured, red blood cells (1) and platelets (2) leak out into the surrounding tissues and a sticky substance called fibrin (3) is produced to help heal the wound.

-  **The damaged vessel** slowly repairs itself and the clot gradually dissolves. Any clots on the surface of the skin turn into scabs, which dry up and fall off.

- **Some people** are not able to produce all the clotting factors. These people will bleed easily unless they receive injections of the missing clotting factor.

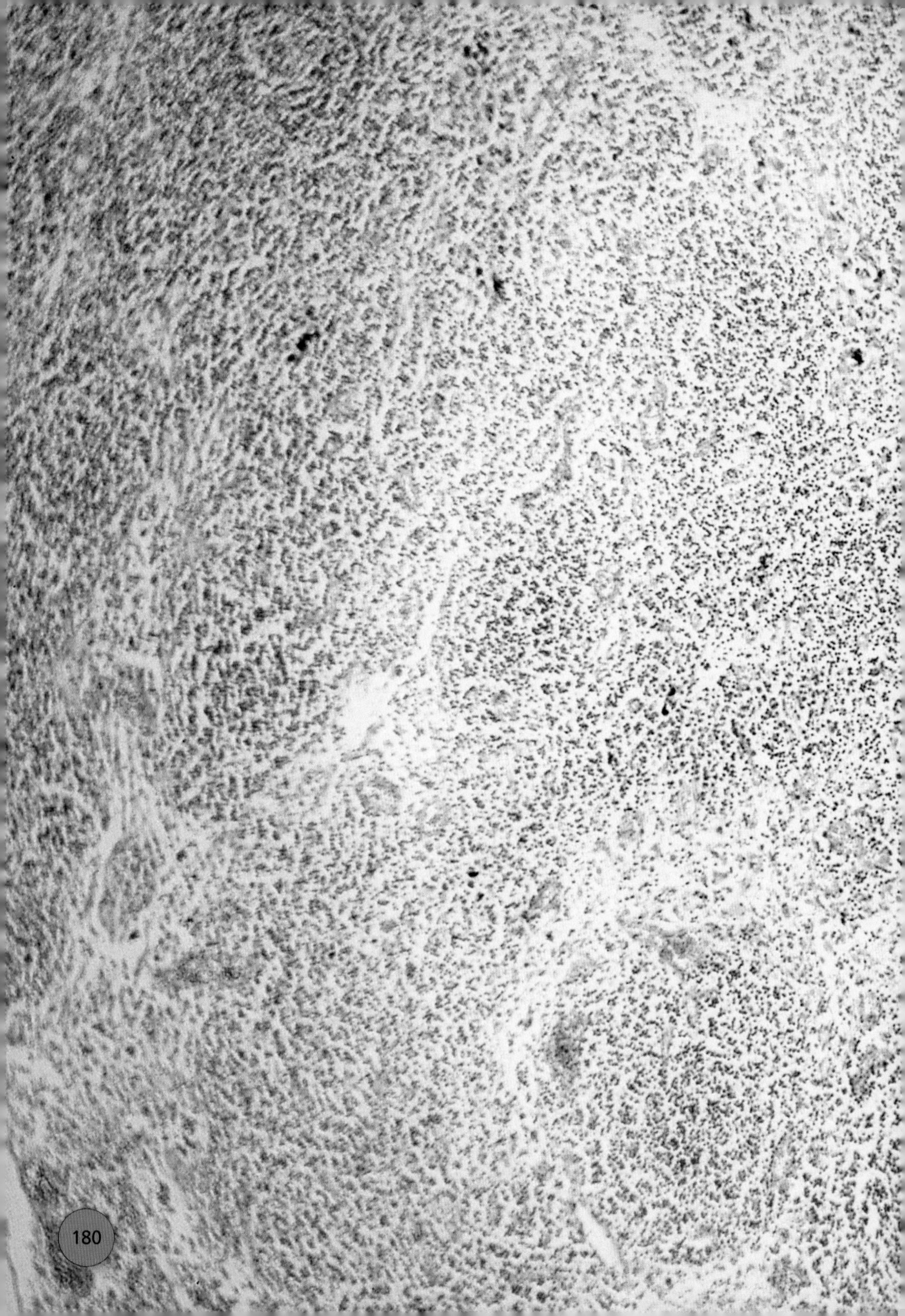

Immune system

The immune system

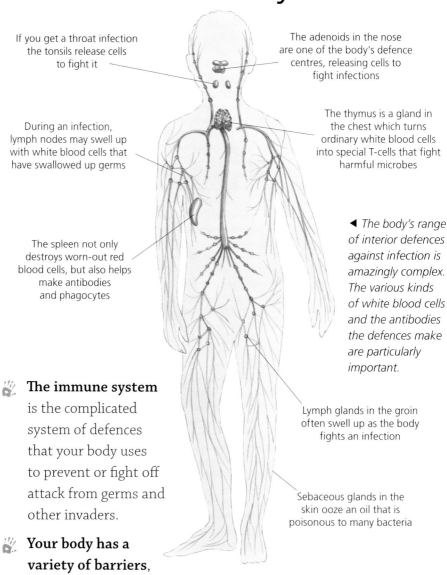

If you get a throat infection the tonsils release cells to fight it

The adenoids in the nose are one of the body's defence centres, releasing cells to fight infections

During an infection, lymph nodes may swell up with white blood cells that have swallowed up germs

The thymus is a gland in the chest which turns ordinary white blood cells into special T-cells that fight harmful microbes

The spleen not only destroys worn-out red blood cells, but also helps make antibodies and phagocytes

◄ The body's range of interior defences against infection is amazingly complex. The various kinds of white blood cells and the antibodies the defences make are particularly important.

Lymph glands in the groin often swell up as the body fights an infection

Sebaceous glands in the skin ooze an oil that is poisonous to many bacteria

The immune system is the complicated system of defences that your body uses to prevent or fight off attack from germs and other invaders.

Your body has a variety of barriers, toxic chemicals and booby traps to prevent germs from entering it. The skin is a barrier that stops many germs getting in, as long as it is not broken.

- **Mucus** is a thick, slimy fluid that coats vulnerable, internal parts of your body such as your stomach and nose. It also acts as a lubricant (oil), making swallowing easier.

- **Mucus lines your airways** and lungs to protect them from smoke particles as well as from germs. Your airways may fill up with mucus when you have a cold, as your body tries to minimize the invasion of airborne germs.

- **Itching, sneezing, coughing and vomiting** are your body's ways of getting rid of unwelcome invaders. Small particles that get trapped in the mucous lining of your airways are wafted out by tiny hairs called cilia.

- **The body** has many specialized cells and chemicals that fight germs that get inside you.

- **Complement** is a mixture of liquid proteins found in the blood that attacks bacteria.

- **Interferons** are proteins that help the body's cells to attack viruses and also stimulate killer cells.

- **Certain white blood cells** are cytotoxic, which means that they are poisonous to invaders.

- **Phagocytes** are big white blood cells that swallow up invaders and then use an enzyme to dissolve them. They are drawn to the site of an infection whenever there is inflammation.

Lymphocytes

- **Lymphocytes** are white blood cells that play a key role in the body's immune system, which targets invading germs.

- **There are two kinds** of lymphocyte – B lymphocytes (B-cells) and T lymphocytes (T-cells).

- **B-cells** develop into plasma cells that make antibodies to attack bacteria such as those that cause cholera, as well as some viruses.

- **T-cells** work against viruses and other micro-organisms that hide inside body cells. T-cells help identify and destroy these invaded cells or their products. They also attack certain bacteria.

- **There are two kinds** of T-cell – killers and helpers.

▼ *A lymph node packed with lymphocytes fighting infection.*

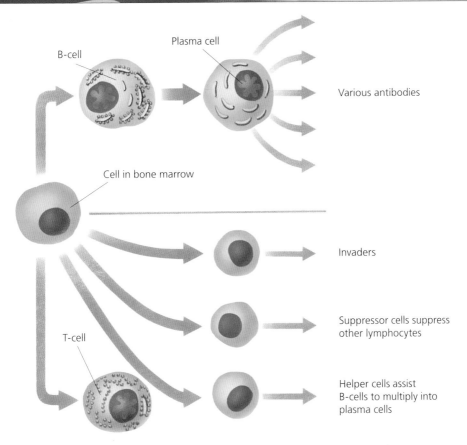

B-cell

Plasma cell

Various antibodies

Cell in bone marrow

Invaders

Suppressor cells suppress other lymphocytes

T-cell

Helper cells assist B-cells to multiply into plasma cells

▲ *Our bodies are constantly under attack from harmful bacteria and viruses. Lymphocytes are key defenders, producing special cells to either identify, alert, suppress or kill.*

Helper T-cells identify invaded cells and send out chemicals called lymphokines as an alarm, telling killer T-cells to multiply.

Invaded cells give themselves away by abnormal proteins on their surface.

Killer T-cells lock on to the cells that have been identified by the helpers, then move in and destroy them.

Some B-cells, called memory B-cells, stay around for a long time, ready for a further attack by the same organism.

Antibodies

- **Antibodies** are tiny proteins that make germs vulnerable to attack by white blood cells called phagocytes.

- **They are produced** by white blood cells derived from B lymphocyctes.

- **There are thousands** of different kinds of B-cell in the blood, each of which produces antibodies against a particular germ.

- **Normally**, only a few B-cells carry a particular antibody. But when an invading germ is detected, the correct B-cell multiplies rapidly to cause the release of antibodies.

- **Invaders are identified** when your body's immune system recognizes proteins on their surface as foreign. Any foreign protein is called an antigen.

- **Your body** was armed from birth with antibodies for germs it had never met. This is called innate immunity.

- **If your body** comes across a germ it has no antibodies for, it quickly makes some. It then leaves memory cells ready to be activated if the germ invades again. This is called acquired immunity.

- **Acquired immunity** means you only suffer once from some infections, such as chickenpox. This is also how vaccination works.

> **DID YOU KNOW?**
> Human beings each generate around ten billion different antibodies.

- **Allergies** are sensitive reactions that happen in your body when too many antibodies are produced, or when they are produced to attack harmless antigens.

- **Autoimmune diseases** are ones in which the body forms antibodies against its own tissue cells.

▼ *Some people's bodies cannot produce antibodies. As a result, they are vulnerable to infections and need special care in a sterile environment.*

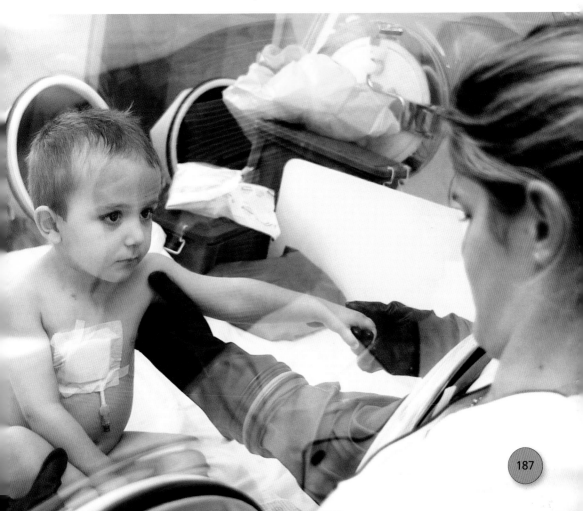

187

Inflammation

Inflammation is the redness, swelling, heat and pain that occurs as the result of an injury or infection.

When the body is damaged, nearby cells release histamines and other chemicals.

▶ *During an inflammatory response, blood vessels widen, causing swelling and redness, while special white blood cells (in purple and green) destroy any infection.*

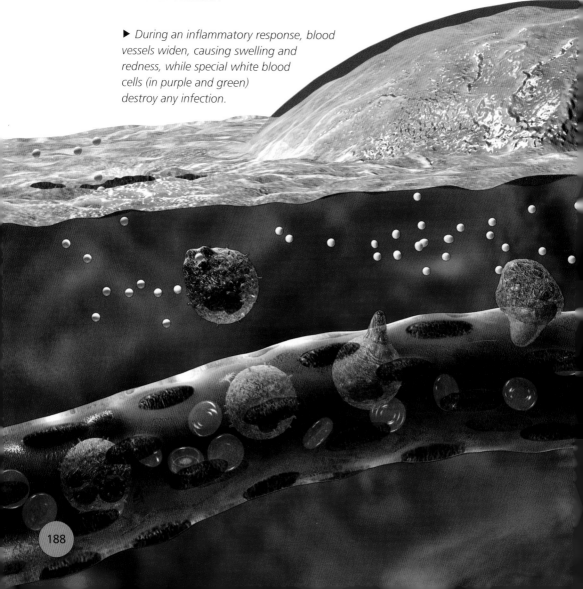

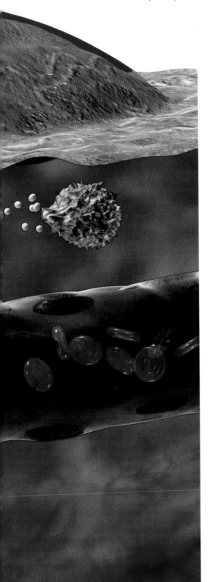

■ **These chemicals** increase blood flow to the area, widening local blood vessels and causing redness and heat.

■ **The chemicals** also attract white blood cells called leucocytes, which destroy any infection.

■ **The leucocytes** also release chemicals that control the inflammation.

■ **Additional fluid** at the injury causes swelling.

■ **The area** may also become painful if nerve endings in the skin are affected.

■ **In some people**, inflammation may occur when there has been no injury. In this case the body will start attacking its own tissues. This is called an autoimmune disease.

■ **Arthritis**, in which joints become swollen, painful and stiff, is an autoimmune disease.

DID YOU KNOW?

Acute inflammation means a sudden reaction that lasts a few days; chronic inflammation means a slower reaction lasting weeks or months.

189

Allergies

Allergies occur when your body produces too many antibodies or produces antibodies against normally harmless antigens. They cause inflammation.

It is not known why some people get allergies and some do not but they often occur in families. You are more likely to get an allergy if your parents, brothers or sisters have an allergy.

Hay fever is an allergy to pollen and causes sneezing, watery and itchy eyes and a runny nose.

Trees, flowers and grasses all produce pollen, and people with hay fever may be allergic to one or several types of pollen.

Some people are allergic to dust. This can cause hay fever-like symptoms such as sneezing all year round.

◄ Flowers are not the only plants that produce pollen – trees, grasses and weeds produce large amounts of pollen too.

- **Pet fur**, nickel in jewellery, chemicals in soaps or perfumes and some foods are all common causes of allergic reactions.

- **Most allergies** are minor but some are life threatening. A severe allergic reaction may cause the mouth and throat to swell up, causing problems with breathing. This is called an anaphylactic reaction.

▶ *During a scratch test, substances that might cause allergies are scratched onto the skin. If a reaction occurs the patient is allergic to that substance.*

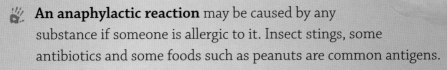

- **An anaphylactic reaction** may be caused by any substance if someone is allergic to it. Insect stings, some antibiotics and some foods such as peanuts are common antigens.

- **Someone with a severe allergy** usually carries drugs that prevent or treat an anaphylactic reaction.

Vaccination

🖐 **Vaccination helps to protect you** against an infectious disease by exposing you to a mild or dead version of the germ in order to make your body build up protection in the form of antibodies.

🖐 **Vaccination is also called immunization**, because it builds up your resistance or immunity to a disease.

🖐 **In passive immunization** you are injected with substances such as antibodies that have already been exposed to the germ. This gives instant but short-lived protection.

🖐 **In active immunization** you are given a killed or otherwise harmless version of the germ. Your body makes the antibodies itself for long-term protection.

▼ *Vaccinations are crucial in many tropical regions where diseases are more widespread.*

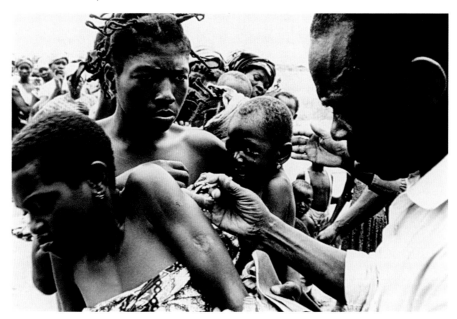

● **Children in many countries** are given a series of vaccinations as they grow up, to protect them against diseases such as diphtheria, tetanus and polio.

● **In cholera**, typhoid, rabies and flu vaccines, the germ in the vaccine is killed to make it harmless.

● **In measles**, mumps, polio and rubella vaccines, the germ is live attenuated – this means that its genes or other parts have been altered in order to make it harmless.

▲ *Diseases such as diphtheria and whooping cough are now rare in many countries thanks to vaccination.*

● **In diphtheria** and tetanus vaccines, the germ's toxins (poisons) are removed to make them harmless.

● **The hepatitis B vaccine** can be prepared by genetic engineering.

● **A new flu vaccine** is developed every year to protect against the current strain of the disease.

193

The lymphatic system

▶ *The lymphatic system is a branching network of little tubes that reaches throughout the body. It drains back to the centre of the body, running into the branches of the superior vena cava, the body's main vein to the heart.*

Drainage back into the blood system

Lymphatics (lymph vessels)

Concentrations of lymph nodes

Lymphatics (lymph vessels)

- **The lymphatic system** is your body's sewer, the network of pipes that drains waste from the cells.

- **It helps** to protect the body against infection by filtering out infectious organisms and helps to keep the amount of fluid in the body stable.

- **The 'pipes'** of the lymphatic system are called lymphatics or lymph vessels.

- **The lymphatics** are filled by a watery liquid called lymph fluid that, along with bacteria and waste chemicals, drains from body tissues such as muscles.

- **The lymphatic system** has no pump, such as the heart, to make it circulate. Instead, lymphatic fluid is circulated as a side effect of the heartbeat and muscle movement.

- **Lymph fluid** drains back into the blood via the body's main vein, the superior vena cava.

- **Valves within the lymphatics** prevent flow backwards along the vessels.

- **The lymphatic system** also consists of clumps of lymph nodes that help filter out infections.

- **The lymphatic system** is not only the lymphatics and lymph nodes, but includes the spleen, the thymus, the tonsils and the adenoids.

- **There are also collections** of lymphatic tissue in the intestines, called Peyer's patches.

- **On average**, at any time about 1–2 litres of lymph fluid circulate in the lymphatics and body tissues.

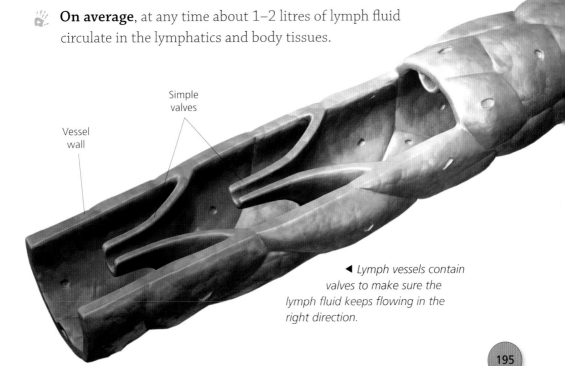

Simple
valves

Vessel
wall

◀ Lymph vessels contain
valves to make sure the
lymph fluid keeps flowing in the
right direction.

195

Lymph nodes

At places in the lymphatic system there are tiny lumps called nodes. These are filters that trap germs that have got into the lymph fluid.

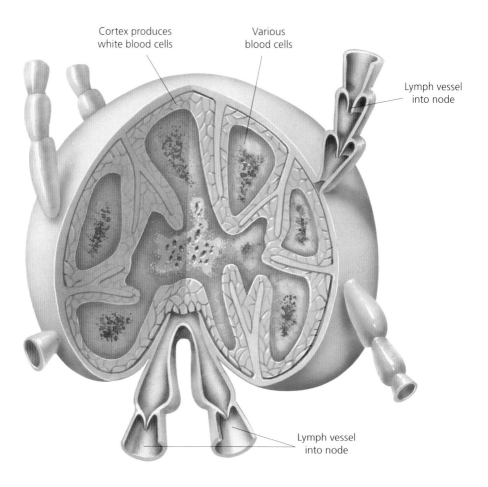

Cortex produces white blood cells

Various blood cells

Lymph vessel into node

Lymph vessel into node

▲ *This shows a cross-section of a lymph node. White blood cells are produced and stored here, and are released through the lymph vessels into the bloodstream.*

- **We have collections** of lymph nodes in our neck, armpits, groin and within our chest and abdomen.

- **Lymph from particular areas** of the body filters into the nearest collection of nodes before passing on through the lymphatic system.

- **Lymph nodes** are covered with fibrous tissue. Inside the node is a network containing white blood cells.

- **As the lymph fluid** passes through the nodes it slows down so that armies of white blood cells called lymphocytes neutralize or destroy germs.

- **When you have a cold** or any other infection, the lymph nodes in your neck or groin, or under your arm, may swell as the lymphocytes fight germs. This is sometimes called 'swollen glands'.

- **Lymphocytes** also destroy cancer cells, which spread through the lymph fluid.

- **The tonsils** and the adenoids are bunches of lymph nodes that swell to help fight ear, nose and throat infections, especially in young children.

- **The adenoids** are at the back of the nose, and the tonsils are at the back of the upper throat.

- **If tonsils or adenoids** swell too much, they are sometimes taken out.

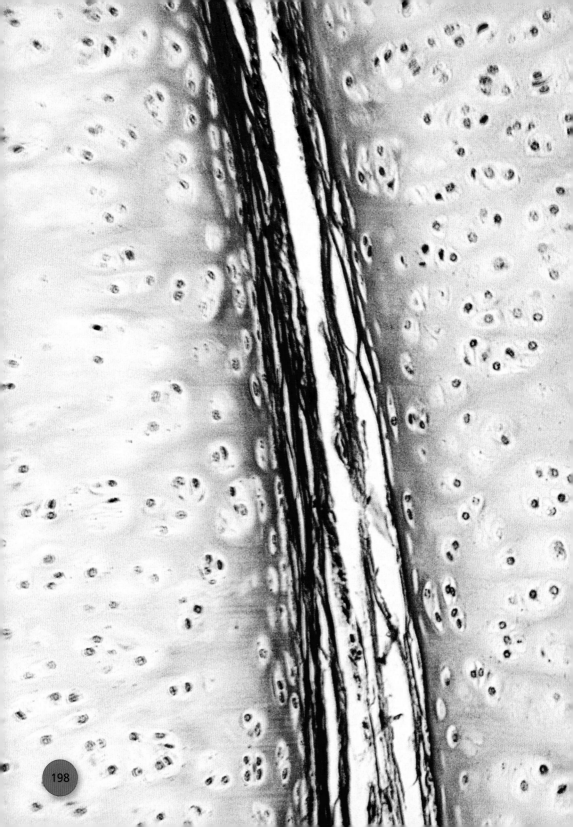

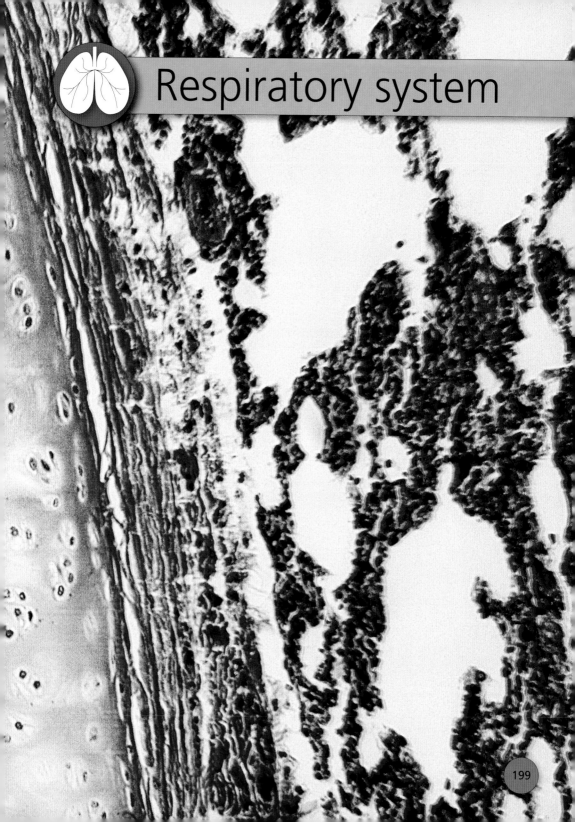

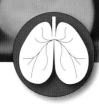

Respiratory system

- **The respiratory system** consists of the airways and the lungs.

- **It is necessary** for gas exchange – exchanging oxygen in the air that we breathe in for carbon dioxide in the air that we breathe out.

- **As air is breathed in** it passes through the nose, down the airways and into the lungs. Breathed out air passes in the opposite direction.

- **The respiratory system** works in partnership with the circulatory system to make sure every cell in the body receives oxygen and is able to get rid of its waste carbon dioxide.

- **The respiratory system** is controlled by the respiratory centre in part of the brain.

- **The respiratory system** takes up most of the space in the chest.

- **The lungs** and the lower part of the airways are enclosed and protected by the ribs.

- **Muscles in the rib cage** contract and relax as we breathe in and out.

- **A large muscle** under the lungs called the diaphragm also contracts and relaxes as we breathe.

DID YOU KNOW?
If you could open up your lungs and lay them out flat, they would cover half a football field.

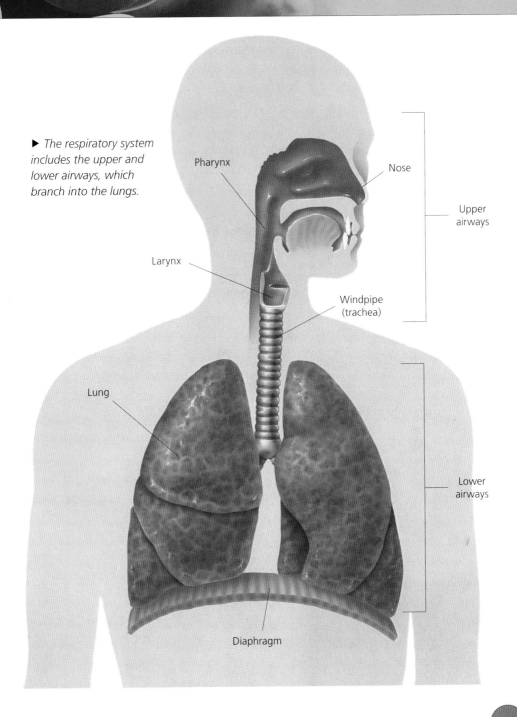

▶ *The respiratory system includes the upper and lower airways, which branch into the lungs.*

Pharynx

Nose

Upper airways

Larynx

Windpipe (trachea)

Lung

Lower airways

Diaphragm

Airways

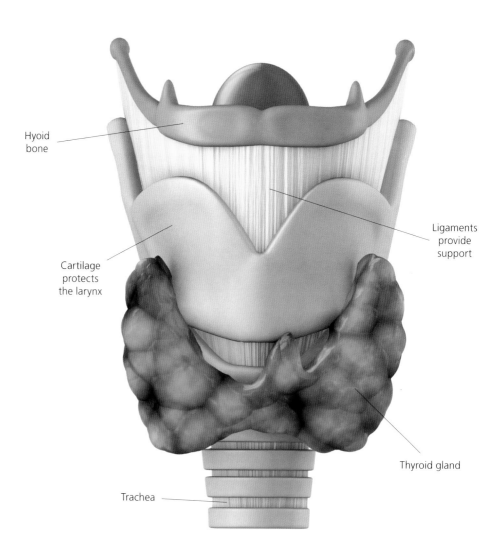

Hyoid bone

Ligaments provide support

Cartilage protects the larynx

Thyroid gland

Trachea

▲ Air travels in through your nose, down the throat and past the larynx, before entering the trachea and travelling on to the lungs.

- **The upper airways** include the nose, the pharynx (throat) the larynx and the trachea (windpipe).

- **The lower airways** include the two main branches into the two lungs (the bronchi) and the small airways of the lungs (bronchioles).

- **Your throat** is the tube that runs down through your neck from the back of your nose and mouth.

- **The throat** is lined by rings of cartilage that keep it open while you breathe in and out.

- **Your throat** branches in two at the bottom. One branch, the oesophagus, takes food to the stomach. The other, the larynx, takes air to the lungs.

- **The two biggest airways** are called bronchi (singular bronchus), and they both branch into smaller airways called bronchioles.

- **The lining** of your airways is protected by a slimy film of mucus that traps dust and other particles to prevent them from getting into the lungs.

- **When you have a cold** your airways produce extra mucus to protect against infection.

- **There are also tiny hairs** called cilia in the lining of the airways. These help clear the lungs of dust and foreign particles.

> **DID YOU KNOW?**
> Your throat is linked to your ears by tubes that open when you swallow to balance air pressure.

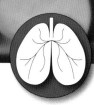

The lungs

- **Your lungs** are a pair of soft, spongy bags inside your chest.

- **The lungs** consist of bunches of minute air sacs called alveoli (singular alveolus).

- **Each bunch** of alveoli is found at the end of a bronchiole – small airways that eventually connect to larger airways and to the throat.

- **Alveoli are surrounded** by a network of tiny blood vessels, and alveoli walls are just one cell thick – thin enough to let oxygen and carbon dioxide seep through them.

- **There are around 300 million** alveoli in your lungs.

▶ Taken through a powerful microscope, this photo of a slice of lung tissue shows a blood vessel and the very thin walls of an alveolus next to it.

Alveoli

Alveolar walls

Capillary walls

Inside capillary

Another capillary

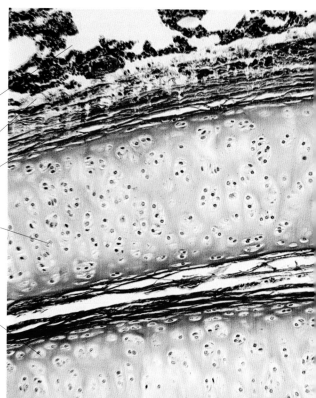

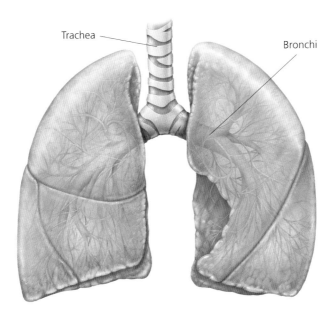

Trachea

Bronchi

▲ *Your trachea divides into two tubes called bronchi. One enters each lung.*

The large surface area of all these alveoli makes it possible for huge quantities of oxygen to seep through into the blood. Equally large amounts of carbon dioxide can seep back into the airways for removal when you breathe out.

If your lungs could be opened up and all the alveoli laid out flat they would cover one half of a tennis court.

People who live at high altitude have a bigger lung capacity than people who live at lower altitudes. This is because the air is thinner at high altitudes.

The alveoli can be damaged by substances such as tar in cigarettes. This makes it harder for oxygen to pass into the blood and causes breathing problems.

It is possible to survive with just one lung if one is damaged or diseased.

205

Breathing

Breathing in   **Breathing out**

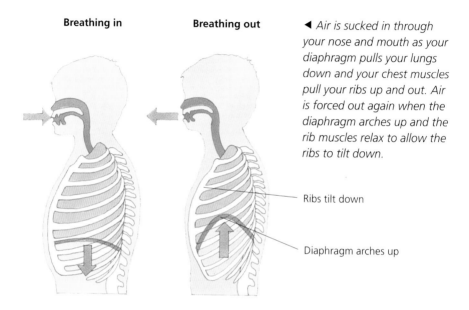

◄ *Air is sucked in through your nose and mouth as your diaphragm pulls your lungs down and your chest muscles pull your ribs up and out. Air is forced out again when the diaphragm arches up and the rib muscles relax to allow the ribs to tilt down.*

Ribs tilt down

Diaphragm arches up

You breathe because every single cell in your body needs a continuous supply of oxygen to burn glucose, the high-energy substance from digested food that cells get from blood.

Scientists call breathing 'respiration'. Cellular respiration is the way that cells use oxygen to burn glucose.

When you breathe in, air rushes in through your nose or mouth, down your windpipe and into the millions of branching airways in your lungs.

On average you breathe in about 15 times a minute. If you run hard, the rate soars to around 80 times a minute.

Newborn babies breathe about 40 times a minute.

- **If you live** to the age of 80, you will have taken well over 600 million breaths.

- **A normal breath** takes in about 0.4 litres of air. A deep breath can take in ten times as much.

- **Your diaphragm** is a dome-shaped sheet of muscle between your chest and stomach, which works with your chest muscles to make you breathe in and out.

- **Scientists call breathing in** 'inhalation' and breathing out 'exhalation'.

- **Breathing is automatic** and controlled by the brain. Although we can make ourselves breathe faster or hold our breath, normally we are not aware of our breathing.

DID YOU KNOW?
Although we need oxygen to live, the air we breathe is about 78 percent nitrogen and only 21 percent oxygen.

▶ When playing a wind or brass instrument, the diaphragm and chest control the air flowing in and out of the lungs.

207

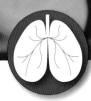

Gas exchange

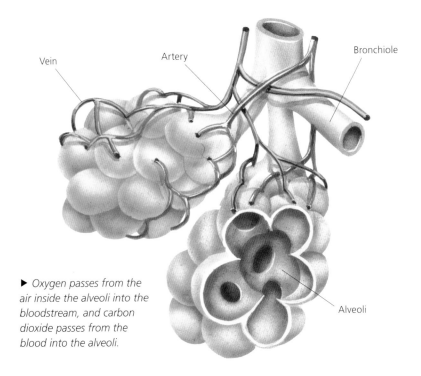

Vein

Artery

Bronchiole

▶ *Oxygen passes from the air inside the alveoli into the bloodstream, and carbon dioxide passes from the blood into the alveoli.*

Alveoli

Gas exchange is the exchange of oxygen and carbon dioxide and takes place throughout the body. All cells need oxygen to survive and need to get rid of waste carbon dioxide.

When you breathe in, air that contains oxygen is taken into your lungs.

In your lungs, the oxygen passes across the thin walls of the alveoli into the capillaries that surround the alveoli.

Once in the capillaries, the oxygen binds with haemoglobin in the red blood cells.

- **The oxygen** is then carried in your blood to your body cells.

- **In the body cells**, oxygen is released from the red blood cells and passes across the capillary walls to be used by the body.

- **At the same time**, waste carbon dioxide from the body cells passes across the capillary walls and dissolves in the plasma in the blood.

- **The carbon dioxide** is then returned by the blood to the lungs.

- **In the lungs**, the carbon dioxide leaves the plasma, crossing through the capillary walls and into the alveoli.

- **The carbon dioxide** is then breathed out.

▼ Scuba divers have to carry tanks of air to allow them to breathe underwater.

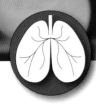

Coughing and hiccups

- **Coughing** is the body's way of removing foreign particles and irritants in the airways and lungs.

- **Irritation of the airways** causes a reflex reaction by stimulating nerves in the airways. These send a signal to the brain that triggers a cough.

- **The irritants** are expelled from the airways, making the distinctive coughing sound.

- **Although a cough** is normally a reflex action, we can also cough voluntarily if we want to.

- **Coughs are usually** caused by an infection, such as a cold. Pollution, cigarette smoke and dust can also cause coughing.

- **Hiccups** are caused when the muscle under the lungs (the diaphragm) contracts suddenly.

- **This sudden contraction** forces air into the lungs. At the same time your throat closes, making the 'hiccup' sound.

- **Some people get hiccups** after eating spicy food or drinking fizzy drinks. Other people get them for no specific reason.

- **Many people** think that a shock, holding their breath or drinking water out of the wrong side of a glass will cure hiccups.

- **The longest recorded attack** of hiccups is thought to be 69 years and five months.

DID YOU KNOW?
When you cough, you expel about 2.5 litres of air from your lungs.

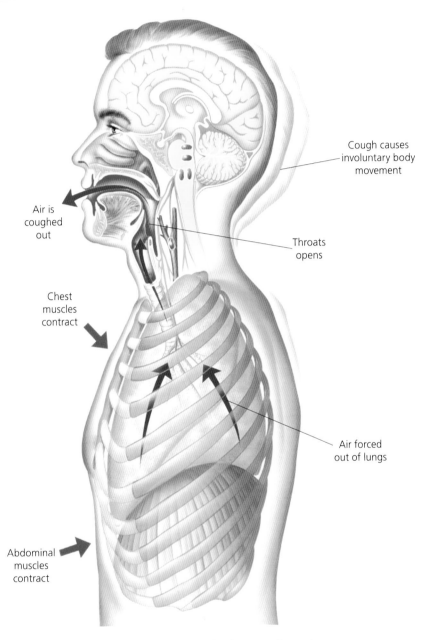

Cough causes
involuntary body
movement

Air is
coughed
out

Throats
opens

Chest
muscles
contract

Air forced
out of lungs

Abdominal
muscles
contract

▲ *During a cough, irritation in your airways stimulates a reflex action, forcing air out of your lungs.*

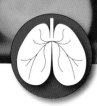

The vocal cords

- **Speaking and singing** depend on the larynx (voice box) in your neck.

- **The larynx** has bands of stretchy fibrous tissue called the vocal cords, which vibrate (shake) as you breathe air out over them.

- **When you are silent**, the vocal cords are relaxed and apart, and air passes freely.

- **When you speak or sing**, the vocal cords tighten across the airway and vibrate to make sounds.

- **The tighter** the vocal cords are stretched, the less air can pass through them, so the higher pitched the sounds you make.

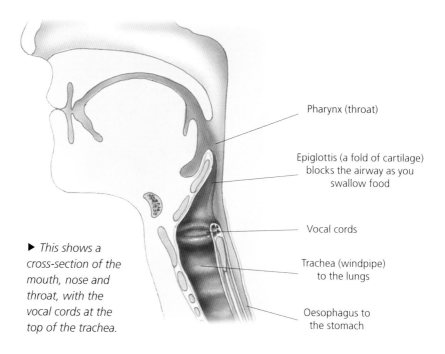

▶ This shows a cross-section of the mouth, nose and throat, with the vocal cords at the top of the trachea.

Pharynx (throat)

Epiglottis (a fold of cartilage) blocks the airway as you swallow food

Vocal cords

Trachea (windpipe) to the lungs

Oesophagus to the stomach

▼ *The vocal cords are soft flaps in the larynx, situated at the base of the throat. Our voices make sounds by vibrating these cords, as shown in the diagram.*

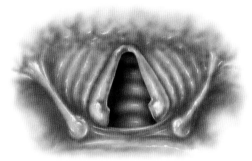 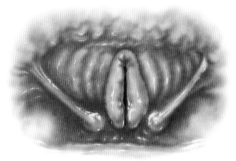

When the cords are apart no sound is made, as air can move freely past them

When the cords are pulled together by tiny muscles, air is forced through a small gap and the cords vibrate to create a sound

- **The basic sound** produced by the vocal cords is a simple 'aah'. But by changing the shape of your mouth, lips and especially your tongue, you can change this simple sound into letters and words.

- **Babies' vocal cords** are just 6 mm long.

- **Women's vocal cords** are about 20 mm long.

- **Men's vocal cords** are about 30 mm long. Because men's cords are longer than women's, they vibrate more slowly and give men deeper voices.

- **Boys' vocal cords** are the same length as girls' until they are teenagers – when they grow longer, making a boy's voice 'break' and get deeper.

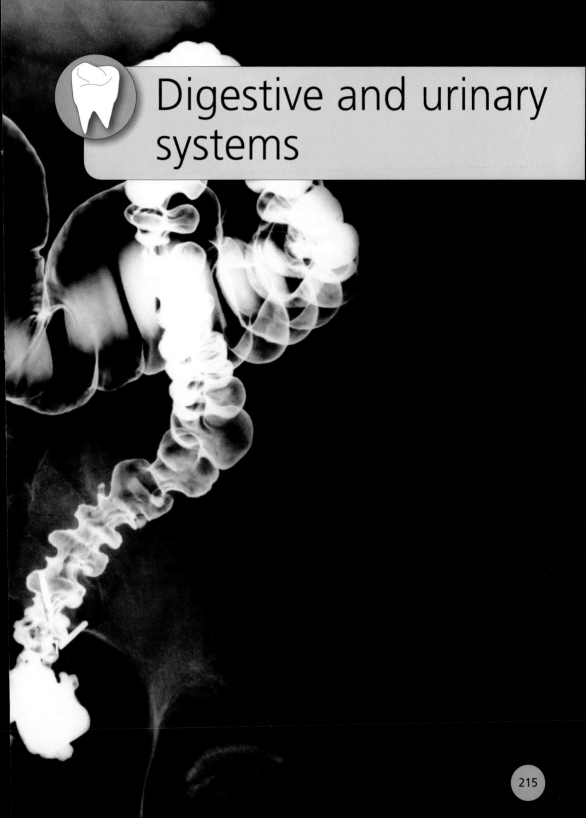

Digestive and urinary systems

Teeth and gums

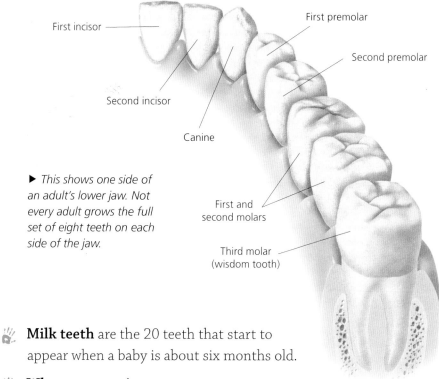

First incisor

First premolar

Second premolar

Second incisor

Canine

▶ *This shows one side of an adult's lower jaw. Not every adult grows the full set of eight teeth on each side of the jaw.*

First and second molars

Third molar (wisdom tooth)

Milk teeth are the 20 teeth that start to appear when a baby is about six months old.

When you are six, you start to grow your 32 adult teeth – 16 in the top row and 16 in the bottom.

Molars are big, strong teeth at the back of the mouth. There are usually six pairs. Their flattish tops are shaped for grinding food.

The four rearmost molars, one in each back corner of each jaw, are the wisdom teeth. These are the last to grow and sometimes they never appear.

Premolars are four pairs of teeth in front of the molars.

Incisors are the four pairs of teeth at the front of your mouth. They have sharp edges for cutting food.

- **Canines** are the two pairs of big, pointed teeth behind the incisors. Their shape is good for tearing food.

- **The enamel** on teeth is the body's hardest substance; dentine inside teeth is softer but still as hard as bone.

- **Teeth sit in sockets** in the jawbones and are held in place by the gums.

- **The gums** are layers of connective tissue that surround each tooth and help to prevent it from damage by infection.

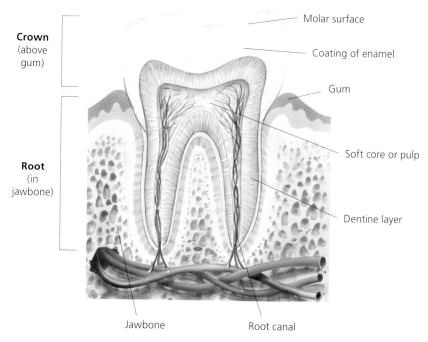

Crown
(above
gum)

Root
(in
jawbone)

Molar surface

Coating of enamel

Gum

Soft core or pulp

Dentine layer

Jawbone

Root canal

▲ Teeth have long roots that slot into sockets in the jawbones, but they sit in a fleshy ridge called the gums. In the centre of each tooth is a living pulp of blood and nerves. Around this is a layer of dentine, then on top of that a tough shield of enamel.

217

Tooth decay

Although teeth are covered in enamel – the hardest substance in the body – they can still be affected by disease.

If teeth are not cleaned regularly, an invisible layer of food and bacteria starts to build up on the surface of the teeth.

This mixture of food and bacteria is called plaque.

Plaque sticks to the teeth and the bacteria feed on the food within the plaque.

As the bacteria feed they produce an acid that starts to erode the enamel.

▶ *Teeth need brushing for several minutes at least twice a day to remove plaque and keep them healthy.*

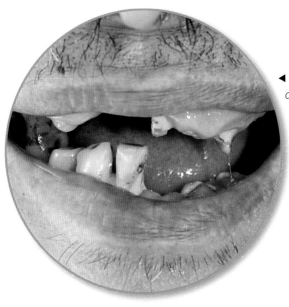

◀ If not treated, tooth decay can damage teeth so badly that they have to be removed.

If the plaque is not removed, eventually it will wear away the enamel and treatment will be needed to fill the hole in the tooth.

Toothaches occur when the acid wears away both the enamel and the dentin underneath and reaches the sensitive nerves in the center of the tooth.

Fillings are usually used to repair holes in teeth caused by tooth decay. If a tooth is badly affected, it may need to be removed.

Plaque is hard to remove, but regular brushing of your teeth and flossing can keep it from building up in the first place.

DID YOU KNOW?

Many countries add fluoride to the water supply to help harden teeth.

219

Digestive system

- **The digestive system** uses the food that we eat and breaks it down into material that body cells can use for energy. It then eliminates what is left over as waste (faeces).

- **The digestive system** consists of the digestive tract and the digestive organs.

- **Your digestive tract** is basically a long, winding tube. It is also called the alimentary canal (gut). It starts at your mouth and ends at your anus.

- **If you could** lay your gut out straight, it would be nearly six times as long as you are tall.

- **Muscular contractions** in the gut move the food that you eat from one end to the other.

- **The digestive system** also consists of several separate organs, including the salivary glands, liver, gall bladder and pancreas.

- **These organs** produce enzymes and chemicals that help in the digestion of food.

- **It takes several days** for food to pass right through the digestive tract.

- **On average**, we digest 30,000 kg of food in a lifetime.

DID YOU KNOW?
In a fully grown adult, the digestive tract is about 7 m long.

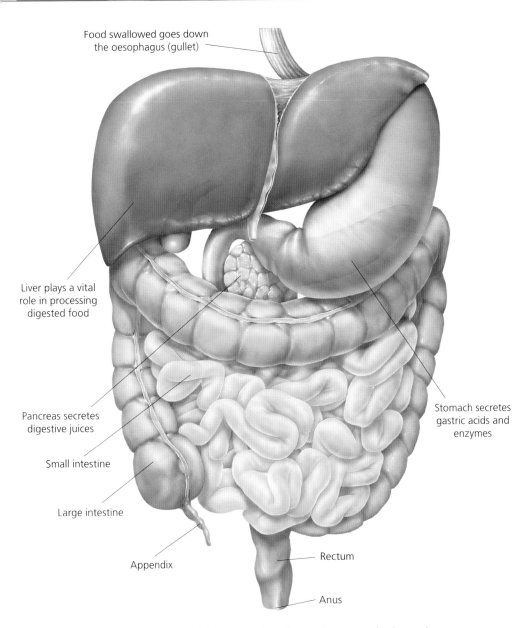

Food swallowed goes down
the oesophagus (gullet)

Liver plays a vital
role in processing
digested food

Pancreas secretes
digestive juices

Small intestine

Large intestine

Appendix

Stomach secretes
gastric acids and
enzymes

Rectum

Anus

▲ The food you eat is broken down into the nutrients your body needs as
it passes down through your oesophagus into your stomach and your small
intestine. Undigested food travels through your large intestine and leaves your
body via your anus.

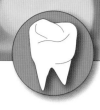

Hunger and appetite

- **Hunger occurs** when the body needs food; appetite is the desire to eat.

- **When the stomach is empty** or the sugar levels in our blood become lower than normal we start to feel hungry.

- **The feeling of hunger** comes from the hypothalamus – a gland in our brain.

- **Hunger often** gives us 'stomach pangs' or makes our stomachs rumble. This is the stomach contracting.

- **The feeling of hunger** usually begins a few hours after we last ate. Stomach pangs usually start around 12 hours after we last ate.

- **The smell and sight** of food may give us an appetite, making us look forward to eating and want to eat.

- **You can be hungry** without having an appetite or a desire to eat, and you can have an appetite or fancy eating something without being hungry.

LESS FILLING	MORE FILLING
Bowl of cornflakes or chocolate cereal	Bowl of porridge
Biscuit	Apple
Packet of crisps	Small bag of popcorn
Cheese on white toasted bread	Beans on brown toasted bread
Can of cola	Glass of milk
Bag of sweets	Pot of yoghurt
Doughnut	Small packet of roasted peanuts
Packet of pretzels	Chocolate and nut bar
Portion of mashed potato	Portion of wholemeal pasta

People who eat when they are not feeling hungry may eat more than they need and become overweight.

Some people prefer not to eat, even when they feel hungry. In extreme cases this can cause illness or be life threatening.

There are many people in the world who do not have enough to eat and go hungry every day.

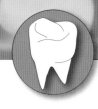

Digestion

- **Digestion is the process** by which your body breaks down the food you eat into substances that it can absorb (take in) and use.

- **Digestion mainly takes place** in the stomach and the small intestine.

- **Food is partly broken down by** mechanical processes, such as chewing and movements in the gut, and partly by enzymes.

- **The food you eat** is softened in your mouth by chewing and by chemicals in your saliva (spit).

- **When you swallow**, food travels down your oesophagus (gullet) into your stomach.

- **The stomach partly digests** the food, turning it into a liquid called chyme. This passes into your small intestine.

- **The chyme** is broken down even more in the small intestine and absorbed through the gut wall into the blood.

- **Food that cannot be digested** in your small intestine passes on into your large intestine. It is then pushed out through your anus as faeces when you go to the toilet.

- **Digestive enzymes** play a vital part in breaking food down so it can be absorbed by the body.

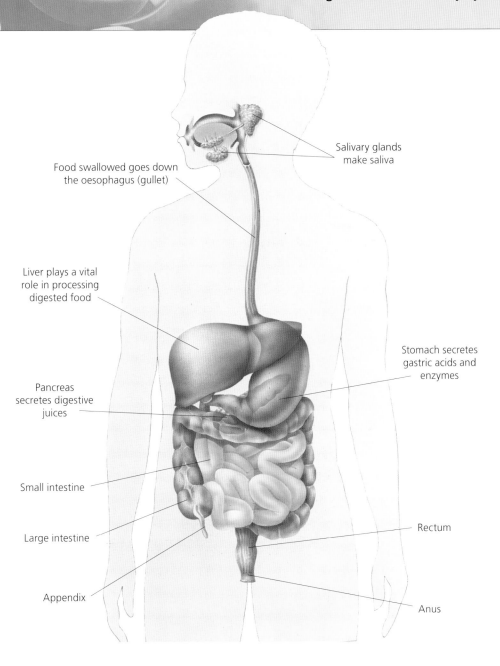

Salivary glands make saliva

Food swallowed goes down the oesophagus (gullet)

Liver plays a vital role in processing digested food

Stomach secretes gastric acids and enzymes

Pancreas secretes digestive juices

Small intestine

Large intestine

Rectum

Appendix

Anus

▲ *The digestive organs almost fill the lower part of the main body, called the abdomen.*

Food processing

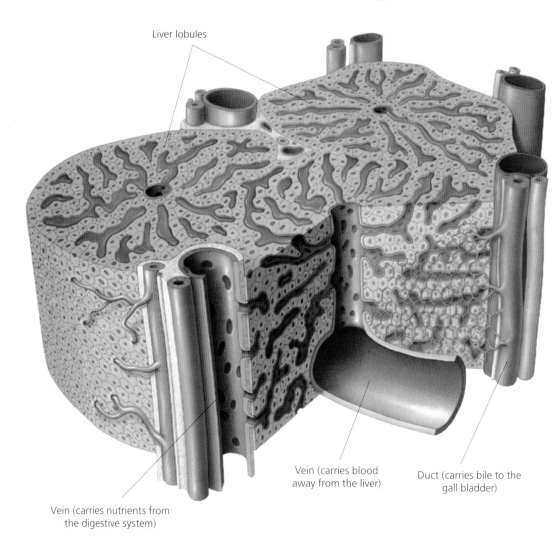

Liver lobules

Vein (carries blood away from the liver)

Duct (carries bile to the gall bladder)

Vein (carries nutrients from the digestive system)

▲ *The liver is the body's main food processing centre. It processes food as we need it and either stores the results or transfers them into the blood to be sent around the body.*

The food we eat consists of complex substances. During digestion, these molecules are broken down into simple nutrients that can be absorbed into the blood.

DID YOU KNOW?
You need some fat in your body – it is used to cushion and insulate your body and is a great energy store.

Carbohydrates are broken down into glucose.

Proteins are broken down into amino acids.

Fats are broken down into fatty acids.

The liver is the main processing centre of the body and also stores many substances for when they are needed.

Glucose is the body's main source of energy and is use for growth and repair. Excess glucose is kept in the liver and muscles as a substance called glycogen.

Glycogen can be turned back into glucose at once if the body suddenly needs energy.

Amino acids are used to make new proteins, which the body uses to repair damage.

Fatty acids are mainly used to make cell walls. Excess fatty acids are stored in fat cells.

Excess amino acids can be converted into fatty acids and also stored in fat cells.

Enzymes

Enzymes are molecules that are mostly protein, and that alter the speed of chemical reactions in living things.

There are thousands of enzymes inside your body – it would not be able to function without them.

Some enzymes need an extra substance, called a coenzyme, to work. Many coenzymes are vitamins.

Most enzymes have names ending in 'ase', such as lygase, protease and lipase.

Pacemaker enzymes play a vital role in controlling your metabolism – the rate at which your body uses energy.

One of the most important enzyme groups is that of the messenger RNAs, which are used as communicators by the nuclei of body cells.

Many enzymes are essential for the digestion of food, including lipase, protease, amylase, and the peptidases. Many of these enzymes are made in the pancreas.

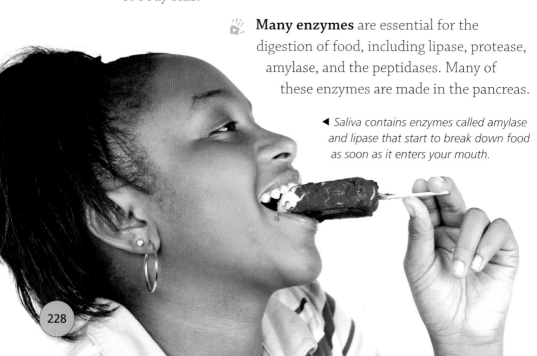

◀ *Saliva contains enzymes called amylase and lipase that start to break down food as soon as it enters your mouth.*

▲ *After you eat a meal, a complex series of enzymes gets to work, breaking food down into simple molecules that can be absorbed into your blood.*

- **Lipase** is released mainly from the pancreas into the alimentary canal (gut) to help break down fat.

- **Amylase** breaks down starches such as those in bread and fruit into simple sugars. There is amylase in saliva and in the stomach.

- **In the gut**, the sugars maltose, sucrose and lactose are broken down by maltase, sucrase and lactase.

Swallowing

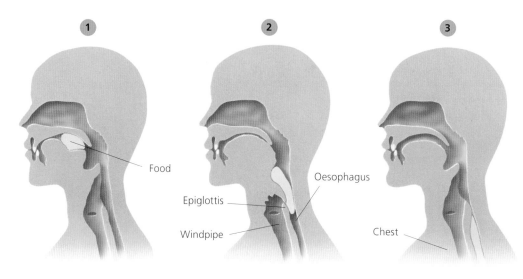

▲ *After chewing (1), food is swallowed into the gullet, or oesophagus (2). This pushes the food powerfully down through the chest (3), past the heart and lungs, into the stomach.*

For digestion to start we need to chew food to make it soft and easy to swallow and digest.

As we chew, the salivary glands around the mouth produce saliva that helps to soften the food so that it is easier to swallow.

Saliva also contains enzymes that start to break down and digest the food.

Swallowing occurs when you push some chewed food towards the back of your mouth.

As you swallow, two involuntary events occur.

- **The soft palate**, which is a flap of tissue at the back of the mouth, is pressed upwards to stop food getting into your nose.

DID YOU KNOW?

Snakes swallow food whole and can swallow prey much larger than themselves.

- **The epiglottis**, which is a flap of cartilage at the top of the windpipe, tilts down over the larynx to stop food entering the airways.

- **The swallowed food** passes the closed off airways and enters the top of the digestive tract – the oesophagus.

- **Once in the oesophagus** the food is pushed towards the stomach by the muscles in the lining of the oesophagus.

- **This muscle contraction** is called peristalsis and occurs all along the digestive tract.

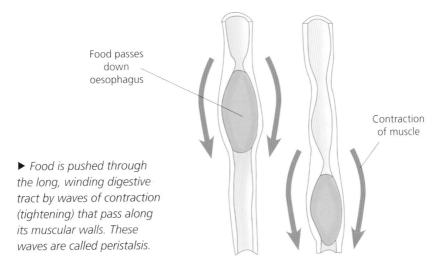

Food passes down oesophagus

Contraction of muscle

▶ Food is pushed through the long, winding digestive tract by waves of contraction (tightening) that pass along its muscular walls. These waves are called peristalsis.

The stomach

Your stomach is a muscular-walled bag that mashes food into a pulp, helped by chemicals called gastric juices.

The word 'gastric' means to do with the stomach.

When empty, your stomach holds barely 0.5 litres, but after a big meal it can stretch to more than 4 litres.

The sight, smell and taste of food all start the production of gastric juices so that by the time the food reaches the stomach it is ready to start digestion.

The stomach lining is protected by mucus so that it does not digest itself.

The gastric juices produced by the stomach contain acid and enzymes that break down proteins.

The stomach also produces a substance called intrinsic factor that is necessary for us to absorb vitamin B12.

The stomach takes up to five hours to mix and digest solid food.

Once the food has been reduced to a liquid, the muscular ring at the exit of the stomach relaxes and the liquid enters the small intestine.

The mixture of churned food and gastric juices that leaves the stomach is called chyme.

Gall bladder

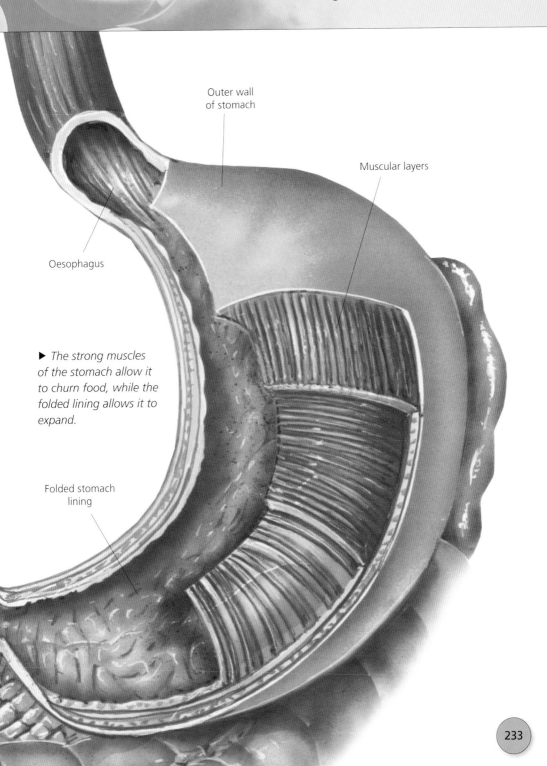

Outer wall
of stomach

Muscular layers

Oesophagus

▶ *The strong muscles
of the stomach allow it
to churn food, while the
folded lining allows it to
expand.*

Folded stomach
lining

233

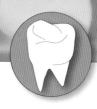

The liver

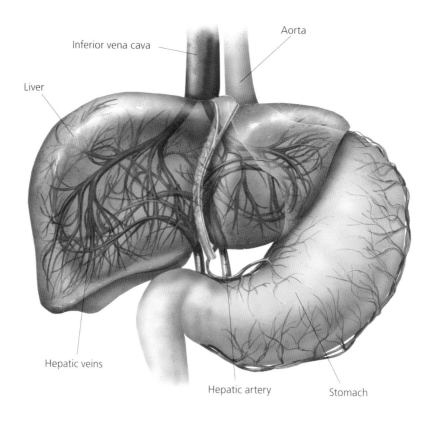

Inferior vena cava

Aorta

Liver

Hepatic veins

Hepatic artery

Stomach

▲ *Located under the ribs and next to the stomach, the liver is well supplied with blood vessels.*

The liver is the body's chemical processing centre.

It is the biggest internal organ, and the word hepatic means 'to do with the liver'.

The prime task of the liver is handling all the nutrients and substances digested from the food you eat and sending them out to your body cells when needed.

- **The liver turns** carbohydrates into glucose, the main energy-giving chemical for body cells.

- **Levels of glucose** in the blood are kept steady by the liver. It does this by releasing more when levels drop, and by storing it as glycogen, a type of starch, when levels rise.

- **The liver** packs off excess food energy to be stored as fat around the body.

- **It also breaks** down proteins and stores vitamins and minerals.

- **Bile** is a yellowish or greenish bitter liquid produced by the liver. It helps dissolve fat as food is digested in the intestines.

- **The liver clears** the blood of old red cells and harmful substances such as alcohol, and makes new plasma.

- **Lobules** are the liver's chemical processing units. They take in unprocessed blood on the outside and dispatch it through a collecting vein.

▶ *Harmful substances such as alcohol, which is found in alcoholic drinks such as wine, are filtered by the liver to help keep the body healthy.*

235

The pancreas

🖐 **The pancreas** is a large, carrot-shaped gland that lies just below your stomach.

🖐 **Two main types** of tissue are contained within the pancreas – one that releases hormones and one that releases pancreatic enzymes.

▼ *The pancreas lies on the right side of the body, tucked into the gut. The microscopic view shows the islets of Langerhans (in purple).*

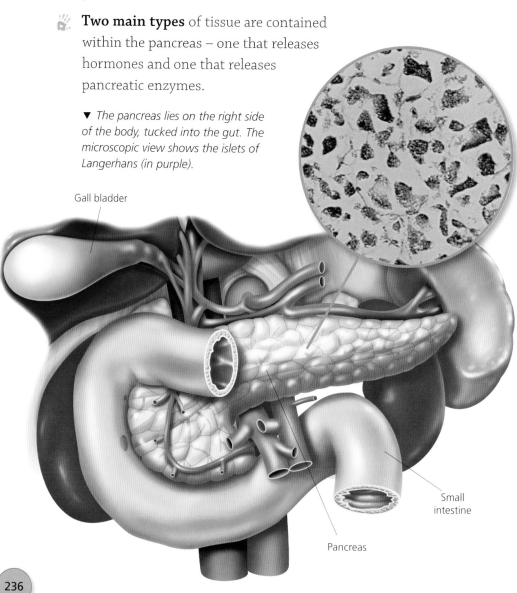

Gall bladder

Small intestine

Pancreas

- **The main type** of tissue consists of thousands of nests of hormone glands called the islets of Langerhans.

- **The islets of Langerhans** release two important hormones: insulin and glucagons.

- **The second type** of tissue is called exocrine tissue and secretes (releases) pancreatic enzymes such as amylase and lipase into the intestine to help digest food.

- **Amylase breaks down** carbohydrates into simple sugars such as maltose, lactose and sucrose.

- **Lipase breaks apart** fat molecules so that they can be absorbed more easily.

- **The pancreatic enzymes** run into the intestine via a pipe called the pancreatic duct, which joins on to the bile duct. This duct also carries bile.

- **The pancreatic enzymes** only start working when they meet other kinds of enzyme in the intestine.

- **The pancreas** also secretes the body's own antacid, sodium bicarbonate, to settle an upset stomach.

DID YOU KNOW?
The pancreas is less than 20 cm long but produces one of the body's most important hormones – insulin.

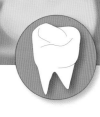

The gall bladder

- **The gall bladder** is a small hollow bag that lies behind the liver and is about 8 cm long.

- **It is connected** to the common bile duct by the cystic duct. The common bile duct connects the liver to the small intestine.

- **The gall bladder stores** a substance called bile, which is made in the liver from waste products.

- **Bile is a bitter yellow** or blue-green liquid that helps to digest fats.

- **The gall bladder** can store about 50 ml of bile.

- **Bile is released** into the small intestine when we eat and helps in the digestion of fats.

- **Bile contains** a fatty material called cholesterol. If there is too much cholesterol in the bile it may start to turn solid and form lumps called gallstones.

Gall bladder

▶ The gall bladder lies just behind the liver and in front of the stomach. It stores bile, a greenish liquid.

- **Gallstones** can reach the size of a golf ball.

- **Some people** with gallstones have their gall bladder removed. You do not need a gall bladder to live.

Stomach

Small intestine

- **Once food** has been broken down into liquid chyme by the stomach it enters the small intestine.

- **Your small intestine** is a 6-m-long tube that is about 2.5 cm wide.

- **It is divided** into three sections, the duodenum, the jejunum and the ileum.

- **The small intestine** is lined with protective mucus to prevent it digesting itself.

- **The lining** also consists of thousands of folds and tiny projections called villi.

- **The villi** give the small intestine a huge surface area. If you could flatten them all out, your small intestine would cover the whole of a tennis court.

- **The first part** of the small intestine, the duodenum, mixes the chyme with enzymes and bile to break it down into molecules small enough to be absorbed.

- **The middle part** of the small intestine, the jejunum, is where food is absorbed. Small molecules such as glucose pass across the villi into the bloodstream, where they can be transported around the body.

◀ *The small intestine takes nutrients and useful substances through its lining, into the body.*

Small intestine

Large intestine

The last part of the small intestine, the ileum, absorbs vitamin B12.

Once all the useful molecules have been absorbed, the remaining liquid passes into the large intestine.

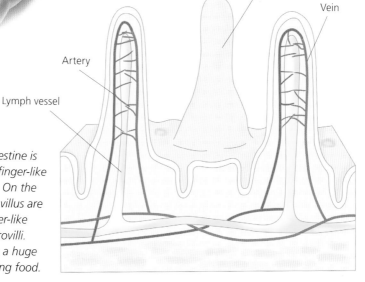

Villus

Vein

Artery

Lymph vessel

▶ *The small intestine is lined with tiny, finger-like folds called villi. On the surface of each villus are even tinier, finger-like folds called microvilli. These folds give a huge area for absorbing food.*

241

Large intestine

- **The first part** of the large intestine is called the cecum. The appendix is attached to the cecum.

- **A small tube** about 10 cm in length, the appendix sometimes gets infected and has to be removed. It has no function in humans.

- **The main part** of the large intestine is the colon, which is almost as long as you are tall.

- **Although it is** much wider than the small intestine, the large intestine has a smaller surface area.

- **Undigested food** in the form of semi-liquid chyme is converted into solid waste by the colon, which absorbs excess water.

- **The colon soaks up** 1.5 litres of water every day.

- **Sodium and chlorine** are also absorbed by the walls of the colon. Bicarbonate and potassium are also removed.

- **Billions of bacteria** live inside the colon and help turn the chyme into faeces. These bacteria are harmless as long as they do not spread to the rest of the body.

- **Bacteria in the colon** make vitamins K and B – as well as smelly gases such as methane and hydrogen sulphide.

- **The muscles** of the colon break the waste food down into segments ready for excretion.

▲ This photograph shows the lumpy lining of the large intestine.
The small purple worm-like objects are 'friendly' bacteria.

Excretion

Digestive excretion is the way your body gets rid of food that it cannot digest.

Undigested food is prepared for excretion in your large intestine or bowel.

Once all the useful molecules and water have been absorbed, the waste that is left leaves the large intestine and collects in your rectum.

This triggers nerve endings in the rectum that make you want to go to the toilet (defecate).

We can control when we defecate. When we decide to go to the toilet the waste is pushed out through our anus.

The anus is a ring of muscle that relaxes to let out the waste.

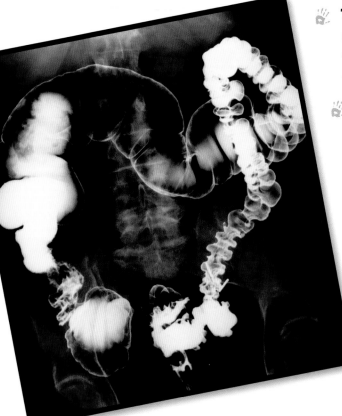

◀ *This is an X-ray of the colon. Patients drink a liquid called barium to enable their doctor to see the colon more clearly and check it is in working order.*

▲ To work well, your bowel needs plenty of fibre – indigestible plant fibres found in food such as beans.

- **If we decide** not to go to the toilet, the waste returns to the large intestine where more water may be absorbed, drying out the waste even more.

- **Fibre is essential** to keep your gut working properly. Fibre is not absorbed so bulks out the waste matter.

- **About two-thirds** of the waste we produce is water.

- **Up to half** of the waste material is friendly bacteria that are found in the gut.

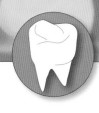

Urinary system

DID YOU KNOW?

During your lifetime you will urinate 45,000 litres – enough to fill a small swimming pool!

- **The urinary system** removes waste products from the blood and from the body.

- **The kidneys** filter the blood to produce urine. This runs down the ureters to collect in the bladder. Urine leaves the body through the urethra.

- **The urinary system** produces about 0.5–2 litres of urine every day.

- **The kidneys** are a pair of bean-shaped organs inside the small of the back.

- **You can survive** with only one kidney.

- **Kidneys were** once thought to be where our conscience, or feelings of right and wrong, were kept.

- **The ureters** are thin tubes that are about 25 cm long. One leads from each kidney to the bladder.

- **The urethra** is a tube that leads from the bladder to the outside of the body.

- **It is longer** in boys and men than in girls and women as it passes through the penis.

▶ The urinary system consists of a pair of kidneys connected to the bladder and a tube (the urethra) to the outside of the body.

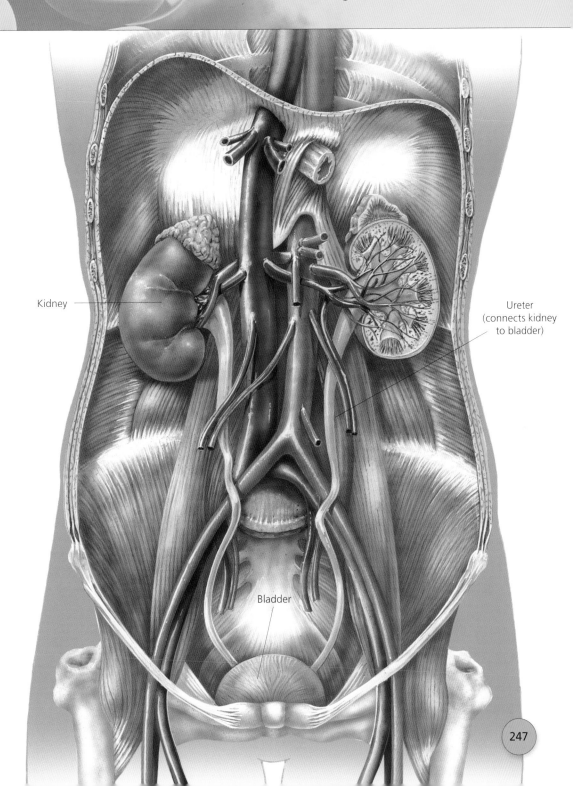

Kidney

Ureter
(connects kidney
to bladder)

Bladder

The kidneys

- **The kidneys** are the body's water control and blood-cleaning plants. They are high-speed filters that draw off water and important substances from the blood. They let unwanted water and waste substances go.

- **About 1.3 litres** of blood are filtered by the kidneys every minute.

- **All the body's blood** flows through the kidneys every ten minutes, so blood is filtered 150 times a day.

- **The kidneys manage** to recycle or save every re-useable substance from the blood. They take 85 litres of water and other blood substances from every 1000 litres of blood, but only let out 0.6 litres as urine.

- **The kidneys** save nearly all the amino acids and glucose from the blood and 70 percent of the salt.

- **Blood entering each kidney** is filtered through a million or more filtration units called nephrons.

- **Each nephron** is an incredibly intricate network of little pipes called convoluted tubules, wrapped around countless tiny capillaries. Useful blood substances are filtered into the tubules, then re-absorbed back into the blood in the capillaries.

- **Blood enters each nephron** through a little cup called the Bowman's capsule via a bundle of capillaries.

- **The kidneys also produce** hormones, including one that helps to keep blood pressure normal.

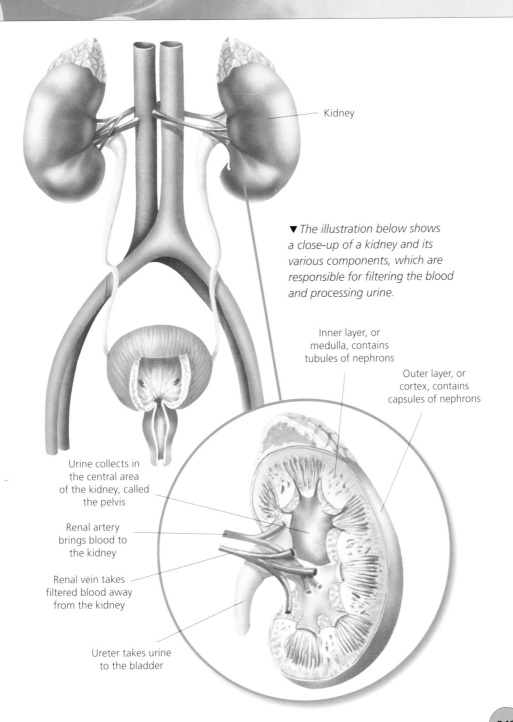

Kidney

▼ *The illustration below shows a close-up of a kidney and its various components, which are responsible for filtering the blood and processing urine.*

Inner layer, or medulla, contains tubules of nephrons

Outer layer, or cortex, contains capsules of nephrons

Urine collects in the central area of the kidney, called the pelvis

Renal artery brings blood to the kidney

Renal vein takes filtered blood away from the kidney

Ureter takes urine to the bladder

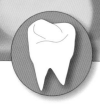

The bladder

- **The bladder** is a hollow, muscular organ that sits deep within the pelvis.

- **It acts as a collection bag** for the urine produced by the kidneys.

- **Urine flows down** the two ureters into the bladder, where it collects until released.

- **The wall of the bladder** is made of smooth muscle and is very folded.

- **As the bladder fills** with urine the folds gradually smooth out so that the bladder can stretch.

- **The bladder** normally holds up to about 600 ml of urine but can stretch to hold up to twice this amount if necessary.

- **As the bladder stretches** it sends nerve signals to the spinal cord and the brain. These tell us we need to empty our bladders (urinate).

- **We start to feel** we need to urinate when the bladder is a quarter full.

- **If we decide** to urinate, we relax the muscle at the base of the bladder and urine passes out through the urethra.

- **At birth**, urination is automatic when the bladder if full. As we grow we learn to control the muscle at the base of the bladder so that we can control when we urinate.

▶ This highly magnified photograph shows the folded lining of the bladder, which flattens out and stretches as the bladder fills with urine.

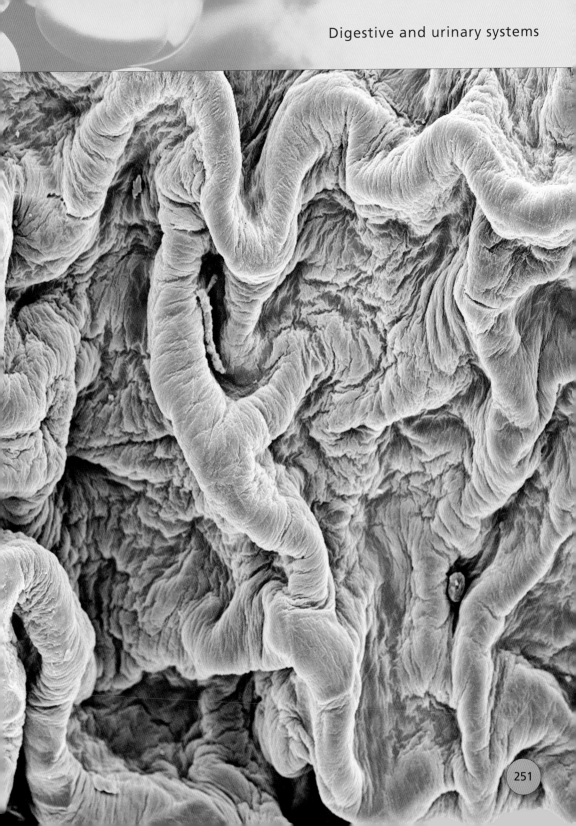

Urine

- **Urine** is one of your body's ways of getting rid of waste.

- **Your kidneys** produce urine, filtering it from your blood.

- **Urine runs from each kidney** down a long tube called the ureter, to a bag called the bladder.

- **Your bladder fills up** with urine over several hours. When it is full, you feel the need to urinate.

- **Urine is mostly water**, but there are substances dissolved in it. These include urea, various salts, creatinine, ammonia and blood wastes.

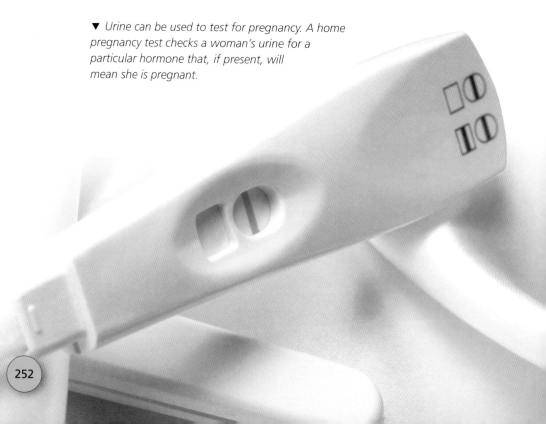

▼ Urine can be used to test for pregnancy. A home pregnancy test checks a woman's urine for a particular hormone that, if present, will mean she is pregnant.

- **Urea** is a substance that is left after the breakdown of amino acids.

- **Urine gets its smell** from substances such as ammonia.

- **Urine gets its colour** from a yellowish blood waste called urochrome. Urochrome is left after proteins are broken down.

- **If you sweat a lot** – perhaps during a fever – your kidneys will let less water go and your urine will be stronger in colour.

- **In some people**, eating beetroot makes their urine go red.

▶ *Doctors use urine dipsticks to check urine for different substances, such as glucose or bacteria. This can help identify various illnesses.*

253

Fluid balance

1000 mL

0.9%
SODIUM CHLORIDE
INJECTION, U

−1
−2
−3
−4
−5
−6
−7
−8
−9

SODIUM
EACH 100 mL CONTAINER
CHLORIDE 900 mg IN 1000 mL
INJECTION. ELECTROLYTE mEq
SODIUM 1.54 mEq; CHLORID
308 mOsmol/LITER (CALC
pH 5.6 (4.5 to 7.0)
ADDITIVES MAY BE INCOMPATIBLE
CONSULT WITH PHARMACIST.
AVAILABLE WHEN INTRODUCING
ADDITIVES USE ASEPTIC TECHNIQUE. MIX
THOROUGHLY AND DO NOT STORE
SINGLE CONTAINER
INTRA USUAL DOSAGE: SEE
INSER NONPYROGENIC. USE
ONLY IS CLEAR AND
CONT MAGED MUST NOT
BE US NNECTIONS.

V. TUBING CHANGED

DRUGS ADDED:

NAME   BOTTLE NO
ROOM NUMBER
USE ON OR BEFORE
REFRIGERATE UNTIL READY TO USE
PREPARED BY: CHECKED BY

- **Your body** balances the amount of water leaving the body to match the amount of water entering the body.

- **We take in water** in our food and drink.

- **We lose water** in urine, faeces, sweat and through moisture in the air that we breathe out.

- **We drink more** if we are thirsty. When the amount of water in the body is low, the brain sends a signal to encourage us to drink.

- **We also become thirsty** if the concentration of certain salts in our body is too high. You will probably feel thirsty after eating a lot of salty food.

- **The amount of water lost** in the urine is controlled by several hormones.

- **The two main hormones** that control fluid balance are aldosterone, which is produced by the adrenal glands that lie just above the kidneys, and anti-diuretic hormone, which is produced by the pituitary gland in the brain.

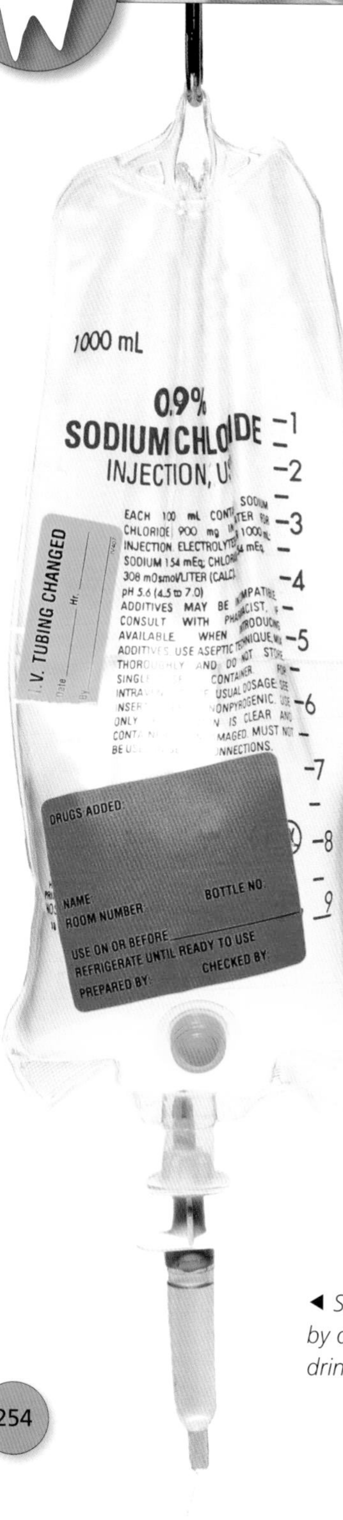

◀ *Saline is a solution of water and salt that is used by doctors to replace fluids in people who cannot drink or have become severely dehydrated.*

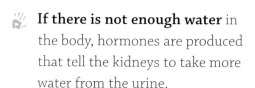

If there is not enough water in the body, hormones are produced that tell the kidneys to take more water from the urine.

If there is too much water in the body, your kidneys will save less water and you will produce more urine.

DID YOU KNOW?
You take in almost as much fluid in the food you eat as you do in the liquid you drink.

▶ *It is important to replace fluid as it is lost, especially if you have been exercising and sweating.*

255

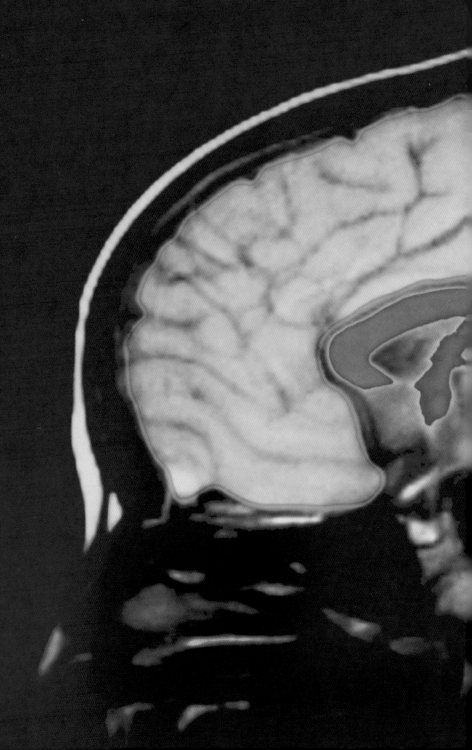

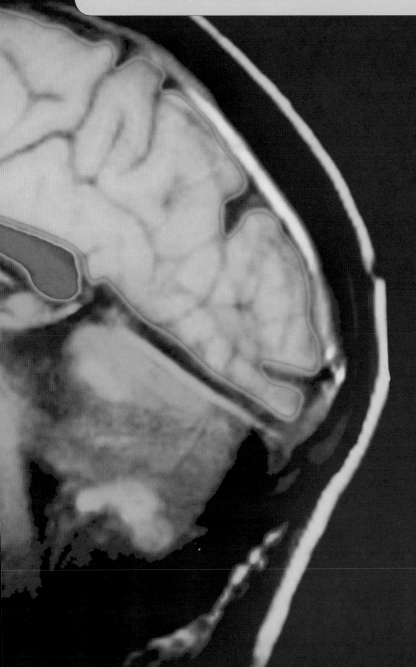

Hormones and metabolism

Hormones

- **Hormones** are the body's chemical messengers, released from stores at times to trigger certain reactions in different parts of the body.

- **Most hormones** are endocrine hormones that are spread around your body in your bloodstream.

- **Each hormone** is a molecule with a certain shape that creates a certain effect on target cells.

- **Hormones are controlled** by feedback systems. This means they are only released when their store gets the right trigger – which may be a chemical in the blood or another hormone.

- **Major hormone sources** include the pituitary gland just below the brain, the thyroid gland, the adrenal glands, the pancreas, a woman's ovaries and a man's testes.

- **Some hormones** only work on certain cells in the body. Others work throughout the body.

- **Endorphins and enkephalins** block or relieve pain.

- **Oestrogen and progesterone** are female sex hormones that control a woman's monthly cycle.

- **Testosterone** is a male sex hormone that controls the workings of a man's sex organs.

- **The word hormone** means 'to excite'.

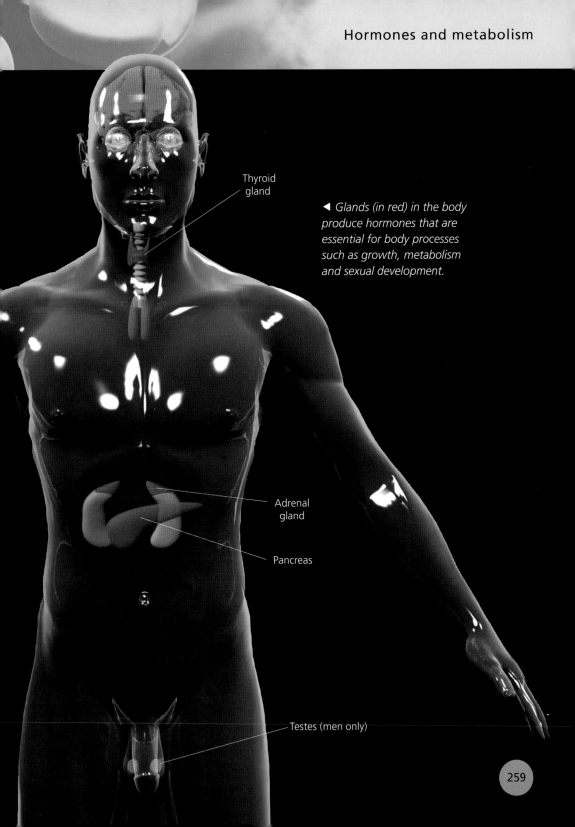

Thyroid
gland

◄ *Glands (in red) in the body
produce hormones that are
essential for body processes
such as growth, metabolism
and sexual development.*

Adrenal
gland

Pancreas

Testes (men only)

Hormonal feedback

- **It is important** that hormone levels are kept just right in the body. If levels become too high or too low they can make us ill.

- **Levels of hormones** in the body are controlled by feedback.

- **A hormone** makes something happen. When it happens, the cells that make that hormone stop producing the hormone.

- **The hormone levels** now fall, meaning the thing it made happen stops. The cells now start producing that hormone again.

- **This type of feedback** is called negative feedback because the event reduces the production of the hormone and hormone levels are kept stable.

- **Occasionally**, an event stimulates cells to produce more of a hormone. The hormone encourages the event and more and more hormone is produced. This is called positive feedback.

- **Positive feedback** occurs in labour when high levels of the hormone oxytocin are needed for birth.

- **Sometimes more than one hormone** controls the same thing. The level of glucose in the blood is controlled by the hormones insulin and glucagon.

- **Some hormones** influence another gland to produce or stop producing hormones.

- **The pituitary gland** influences many of the glands in the body and is sometimes called the master gland.

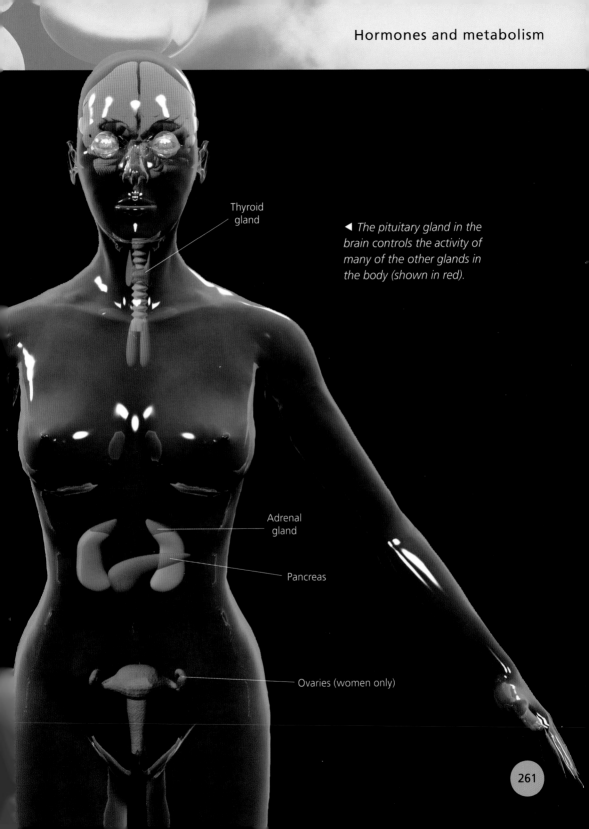

Thyroid
gland

Adrenal
gland

Pancreas

Ovaries (women only)

◄ *The pituitary gland in the
brain controls the activity of
many of the other glands in
the body (shown in red).*

The pituitary gland

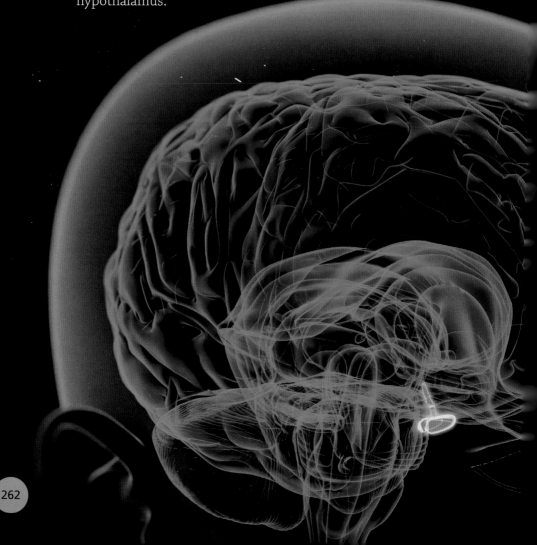

- **The pituitary gland** lies at the base of the brain and is linked to the part of the brain called the hypothalamus.

- **It produces** many important hormones, some of which act on other glands to make them produce hormones.

- **The production** of pituitary gland hormones is controlled by the hypothalamus.

- **Growth hormone** encourages growth in children and teenagers, makes bones stronger and helps build muscles.

- **ACTH**, or adrenocorticotrophic hormone, stimulates the adrenal glands to produce other hormones.

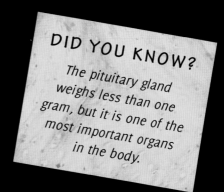

DID YOU KNOW?
The pituitary gland weighs less than one gram, but it is one of the most important organs in the body.

- **Thyroid-stimulating hormone** makes the thyroid gland produce thyroid hormones.

- **Anti-diuretic hormone** acts on the kidneys to reduce the amount of water in the urine.

- **Follicle-stimulating hormone** and luteinizing hormone are involved in reproduction in both men and women. They also affect the production of sex hormones.

- **Oxytocin encourages labour** and makes the breasts release milk when breastfeeding a baby.

- **Prolactin** is produced during and after pregnancy and helps the breasts produce milk for breastfeeding.

◄ *The red, pea-sized object at the base of the brain is the pituitary gland, which produces many hormones that act on other glands throughout the body.*

263

The thyroid gland

The thyroid is a small gland about the size of two joined cherries. It is situated at the front of your neck, just below the larynx.

The thyroid secretes (releases) three important hormones – tri-iodothyronine (T3), thyroxine (T4) and calcitonin.

The thyroid hormones affect how energetic you are by controlling your metabolic rate. Your metabolic rate is the rate at which your body cells use glucose.

T3 and T4 control metabolic rate by circulating into the blood and stimulating cells to convert more glucose.

If the thyroid sends out too little T3 and T4, you get cold and tired, your skin gets dry and you put on weight.

Thyroid gland

◀ The thyroid gland lies in front of your windpipe and produces hormones that help to control energy levels in your body.

▲ *The thyroid is part of your energy control system and tells your body cells to work faster or slower when you need them to.*

DID YOU KNOW?
The thyroid hormone calcitonin controls the amount of calcium in the blood.

The amount of T3 and T4 sent out by the thyroid depends on how much thyroid-stimulating hormone is sent to it from the pituitary gland.

If the levels of T3 and T4 in the blood drop, the pituitary gland sends out extra thyroid-stimulating hormone to tell the thyroid to produce more.

There are four parathyroid glands at the back of the thyroid. These produce parathyroid hormone, which helps control the amount of calcium in the body.

265

The adrenal glands

- **You have two adrenal glands** – one just above each kidney. Each gland is divided into two parts.

- **The outer layer** of the adrenal gland is called the cortex and produces corticosteroid hormones.

- **These corticosteroid hormones** include cortisol, aldosterone and sex hormones.

- **Cortisol increases** the level of glucose in the body and increases blood pressure, helping the body react the stress.

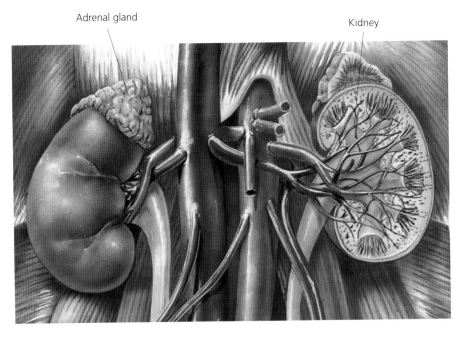

Adrenal gland   Kidney

▲ The adrenal glands sit on the top of each kidney and produce several hormones that help you react to events around you.

Aldosterone helps the kidneys regulate the levels of sodium and potassium in the body.

The sex hormones include testosterone, which stimulates the development of the male reproductive system.

The production of corticosteroid hormones is controlled by a hormone produced by the pituitary gland.

The inner layer of the adrenal glands is called the medulla and it produces adrenaline and noradrenaline.

Adrenaline and noradrenaline get your body ready for action by increasing your heart and breathing rates and the blood flow to your muscles. This is called the 'fight or flight' response.

Adrenaline and noradrenaline are produced when you are threatened or under stress. You may notice the flight or flight response if you have to do something stressful at school, such as making a speech.

▼ *Adrenaline boosts your heartbeat and breathing during exciting or stressful moments.*

267

Pancreatic hormones

The pancreas secretes enzymes that aid digestion. It also releases hormones that regulate blood sugar levels.

These hormones are called insulin and glucagon.

Both are released from clusters of cells in the pancreas called the islets of Langerhans.

Insulin lowers levels of glucose in the blood by encouraging cells to take up glucose from the bloodstream, storing it as fat or as glycogen in the liver and muscle.

Glucagon raises levels of glucose in the blood by encouraging the liver to convert glycogen back into glucose and release it into the bloodstream.

▼ *People with diabetes need to test their blood sugar levels regularly to make sure they are at normal levels.*

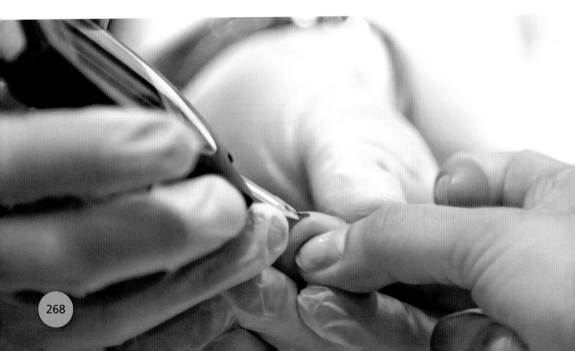

▲ *When we eat sugary food, hormones are released by the pancreas to help keep our blood sugar levels regular.*

Insulin and glucagons balance each other out so that your blood sugar levels are always stable.

Diabetics suffer from the condition diabetes. Some diabetics produce little or no insulin in their pancreas. Others have become resistant to insulin so that their body no longer reacts as it should.

This means that their blood sugar levels are often too high and cells cannot get the energy they need.

Some diabetics are able to control their blood sugar levels by being very careful about what they eat. However many have to control their blood glucose by injecting insulin.

Sex hormones

- **The sexual development** of girls and boys depends on the sex hormones.

- **Sex hormones control** the development of primary and secondary sexual characteristics, and regulate all sex-related processes such as sperm and egg production.

- **Primary sexual characteristics** are the development of the major sexual organs, in particular the genitals.

- **Secondary sexual characteristics** are other differences between the sexes, such as men's beards.

- **There are three** main types of sex hormone – androgens, oestrogen and progesterone.

- **Androgens** are male hormones such as testosterone. They make a boy's body develop features such as a beard, deepen his voice and make his penis grow.

- **Oestrogen** is the female hormone made mainly in the ovaries. It not only makes a girl's sexual organs develop, but also controls her monthly menstrual cycle.

- **Progesterone** is the female hormone that prepares a girl's uterus (womb) for pregnancy every month.

- **Some contraceptive pills** have oestrogen in them to prevent the ovaries releasing their egg cells.

◀ Sex hormones are necessary for normal sexual development in both boys and girls.

271

Male reproductive system

- **The male reproductive system** consists of the penis, scrotum and the two testes (singular, testis).

- **A boy's or man's reproductive system** is where his body creates the sperm cells that combine with a female egg cell to create a new human life.

- **The testes and scrotum** hang outside the body where it is cooler, because this improves sperm production.

- **Sperm cells** look like microscopically tiny tadpoles. They are made in the testes, which are inside the scrotum.

- **At 15**, a boy's testes can make 200 million sperm a day.

- **Sperm leave the testes** via the epididymis – a thin, coiled tube, about 6 m long.

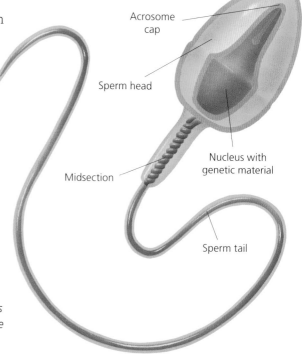

Acrosome cap

Sperm head

Midsection

Nucleus with genetic material

Sperm tail

▶ A mature sperm cell consists of a head, where the genetic information is stored, a midsection and a tadpole-like tail, which allows it to swim rapidly towards the female egg cell.

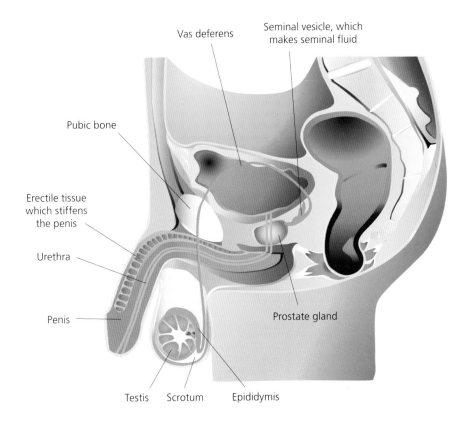

▲ *This is a side view of the inside of the male reproductive organs.*

- **The epididymis** connects to another tube called the vas deferens.

- **Glands called seminal vesicles** lie alongside the vas deferens and add fluids and nutrients to the sperm.

- **The male reproductive system** also produces sex hormones that are needed for the production of sperm and for a boy to develop at puberty.

- **The male sex hormone** testosterone also stimulates bone and muscle growth.

Female reproductive system

- **The female reproductive system** consists of the uterus (womb) fallopian tubes, ovaries, cervix and vagina.

- **A woman's reproductive system** is where her body stores, releases and nurtures the egg cells (ova – singular, ovum) that create a new human life when joined with a male sperm cell.

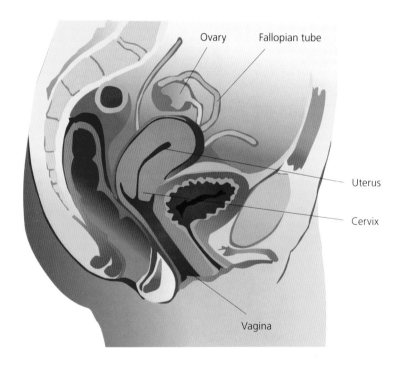

Ovary   Fallopian tube

Uterus

Cervix

Vagina

▲ *This is a side view of the inside of a female reproductive system, showing the two ovaries and fallopian tubes, which join to the uterus.*

- **All the egg cells** are stored from birth in the ovaries – two egg-shaped glands inside the pelvic region. Each egg is stored in a tiny sac called a follicle.

- **Baby girls** are born with around 150,000 egg cells that will gradually mature from puberty onwards.

- **Eggs are gradually released** from the ovaries until menopause, which usually occurs when a woman is between 45 and 55 years of age.

- **The ovaries** sit just near the end of the fallopian tubes. Eggs travel down these tubes to the uterus.

- **The uterus** is a hollow muscular organ that can stretch to fit a baby as it grows in the womb.

- **The entrance** to the uterus is called the cervix. This is usually held closed but opens during labour to let the baby be born.

- **The canal** from the uterus to the outside of the body is called the vagina. It has muscular walls that can stretch during labour.

- **The female reproductive system** also produces sex hormones that are needed for the menstrual cycle and for a girl to develop at puberty. At menopause, the levels of these sex hormones fall.

The menstrual cycle

- **The menstrual cycle** is the way in which a woman's body prepares for pregnancy.

- **It takes place** about every four weeks between puberty and menopause, when hormone levels suddenly drop.

- **Eggs in the ovaries** are stored in follicles. A monthly menstrual cycle starts when follicle-stimulating hormone (FSH) is sent by the pituitary gland in the brain to spur one of the follicles to grow.

- **As the follicle grows**, it releases the sex hormone oestrogen. Oestrogen makes the lining of the uterus (womb) thicken.

- **Halfway through** the menstrual cycle, luteinizing hormone from the pituitary gland encourages the follicle to rupture, and the egg is released.

- **The empty egg** follicle starts to produce progesterone, which also thickens the lining of the uterus.

- **One egg cell** is released every menstrual cycle by one of the ovaries.

- **The egg travels down** the fallopian tubes to the uterus.

- **If the egg is not fertilized**, it is shed with the womb lining in a flow of blood from the vagina. This shedding is called a menstrual period.

- **Menstrual periods** usually occur about every 28 days and last between three and seven days, but vary from woman to woman.

DID YOU KNOW?

If a group of women live together for a long time, it is thought that their menstrual cycles become synchronised.

▼ *This magnified photograph shows an egg bursting from a follicle in the ovary.*

Breasts

- **The breasts** are a mixture of gland tissue and fat, with connective tissue to hold them in shape.

- **They attach** to the main muscle on the chest wall and are held in place by ligaments.

- **Breasts develop** in girls at puberty, usually between the ages of about 11 and 14.

- **The main function** of the breast is to provide milk for newborn babies.

- **During pregnancy**, hormones make a woman's breasts grow and milk glands increase in number and become able to produce milk.

- **Breasts can produce** about one litre of milk a day after a baby is born.

Milk producing glands

◄ Breasts consist of a mixture of fat, connective tissue and glands that produce milk, arranged in four quadrants.

Breastfeeding is best for a baby because breast milk contains natural antibodies that help to protect the baby against infection.

Breasts can become painful just before and during a menstrual period. This is because of changes in hormone levels.

Breasts can change shape and size, especially if a woman gains or loses a lot of weight.

Breast cancer is one of the most common causes of death from cancer in women.

Solid lump in breast

▶ *An X-ray of the breast is called a mammogram. This X-ray shows a solid area, which may be caused by a lump of connective tissue or by cancer.*

279

Reproduction

- **Reproduction occurs** when a developed egg meets a sperm cell during sexual intercourse.

- **When an egg is ripe**, it slides down a duct called a fallopian tube.

- **If a man and woman** have sexual intercourse, the penis is stimulated. Sperm are driven into a tube called the vas deferens and mix with a liquid called seminal fluid to make semen.

- **Semen shoots through** the urethra (the tube inside the penis through which males urinate) and is ejaculated into the female's vagina.

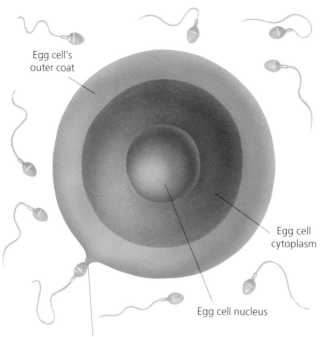

Egg cell's outer coat

Egg cell cytoplasm

Egg cell nucleus

Sperm cell about to fertilize egg cell

◄ The female egg cell passes along the woman's fallopian tube. At fertilization, tiny sperm cells swarm around the egg until one sperm manages to push its head on to the surface of the egg. The sperm head and egg membrane join, and fertilization takes place.

► Reproduction relies on a complex series of events that must be timed exactly. Even then, there is only a chance that a baby will be created.

The sperm from the man's penis may swim up the woman's vagina, enter her womb and fertilize the egg in the fallopian tube.

Although millions of sperm cells are usually released, only one is needed to fertilize the egg.

If the egg is fertilized, the womb lining goes on thickening ready for pregnancy, and the egg begins to develop.

The egg implants into the thick lining of the womb and carries on developing. It is now an embryo or tiny baby.

It takes up to a week for the fertilized egg to implant into the womb. By this time it consists of hundreds of cells.

Rarely, two eggs may be released at the same time and each may be fertilized by a sperm cell. If both implant into the womb, twins develop.

Pregnancy

Pregnancy begins when a woman's ovum (egg cell) is fertilized by a man's sperm cell. Usually this happens after sexual intercourse, but it can begin in a laboratory.

When a woman becomes pregnant her monthly menstrual periods stop. Tests on her urine show whether she is pregnant.

During pregnancy, the fertilized egg divides again and again to grow rapidly – first to an embryo (the first eight weeks), and then to a foetus (from eight weeks until birth).

Unlike an embryo, a foetus has grown legs and arms, as well as internal organs such as a heart.

Pregnancy lasts nine months, and the time is divided into three trimesters (periods of about 12 weeks).

The foetus lies cushioned in its mother's uterus (womb) in a bag of fluid called the amniotic sac.

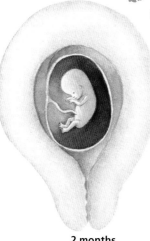

2 months

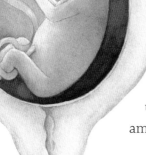

3 months

5 months

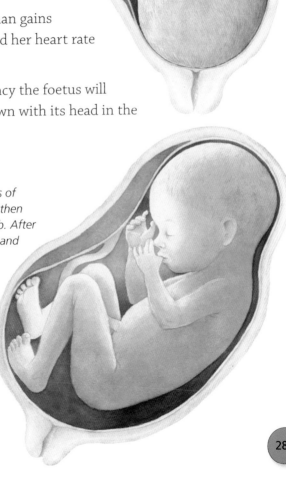

9 months

- **The mother's blood** passes food and oxygen to the foetus via the placenta, also known as the afterbirth.

- **The umbilical cord** runs between the foetus and the placenta, carrying blood between them.

- **During pregnancy** a woman gains 30 percent more blood, and her heart rate goes up.

- **By the end** of the pregnancy the foetus will usually be lying upside down with its head in the pelvis ready to be born.

◄► *These are the various stages of development of an embryo and then foetus inside the mother's womb. After fertilization, the egg cell divides and develops into an embryo. After eight weeks, the embryo is called a foetus.*

7 months

283

Birth

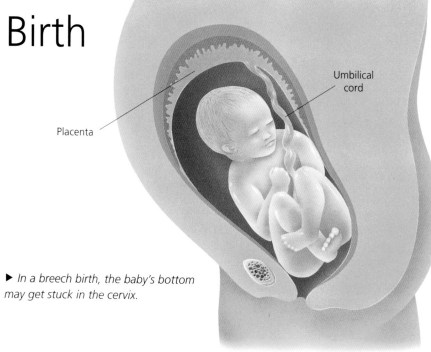

Placenta

Umbilical cord

▶ *In a breech birth, the baby's bottom may get stuck in the cervix.*

Babies are usually born 38–42 weeks after the mother becomes pregnant.

A few days or weeks before a baby is born, it usually turns in the uterus (womb) so its head is pointing down towards the mother's birth canal (her cervix and vagina).

Birth begins as the mother goes into labour – when the womb muscles begin a rhythm of contracting (tightening) and relaxing in order to push the baby out through the birth canal.

There are three stages of labour. In the first, the womb muscles begin to contract or squeeze, bursting the bag of fluid around the baby. This is called breaking the waters.

In the second stage of labour, the baby is pushed out through the birth canal, usually head first.

- **In the third stage** of labour, the placenta, which passed oxygen and nutrients from the mother's blood, is shed and comes out through the birth canal.

- **The umbilical cord** is the baby's lifeline to its mother. It is cut after birth.

- **A premature baby** is one born before it is fully developed.

- **A miscarriage** is when an embryo or foetus is 'born' before it has developed sufficiently to survive independently.

- **A Caesarian section** is an operation that happens when a baby cannot be born through the birth canal and emerges from the womb through a surgical cut made in the mother's belly.

▼ *Babies that weigh below 2.4 kg are known as premature, and are nursed in special care units.*

Babies

- **Newborn babies** usually weigh 3–4 kg and are about 50 cm in length.

- **A baby's head** is three-quarters of the size it will be as an adult – and a quarter of its total body height.

- **The bones** of a baby's skeleton are fairly soft, to allow for growth. They harden over time.

- **There are two gaps** called fontanelles between the bones of a baby's skull, where there is only membrane (a 'skin' of thin tissue), not bone. The gaps close and the bones join together by about 18 months.

- **Babies have a gland** called the thymus gland in the centre of their chest. It is important in providing immunity but gradually shrinks after you reach puberty.

- **A baby** has a very developed sense of taste, with taste buds all over the inside of its mouth.

- **Babies have** a much stronger sense of smell than adults – perhaps to help them find their mothers.

- **Their eyesight** is poor to begin with and their eyelids are usually puffy.

- **A baby is born** with primitive reflexes (things it does automatically) such as grasping or sucking a finger.

- **A baby seems to learn** to control its body in stages, starting first with its head, then moving on to its arms and legs.

▼ *A newborn baby cannot hold up its head, so it must always be supported.*

Growth

- **Growth is controlled** by hormones, especially growth hormone, which is produced by the pituitary gland.

- **Your body proportions** change as you grow. A baby's legs only make up about one-quarter of its length, but by the time we are grown up our legs make up almost half of our height.

- **A baby's body weight** usually triples in the first year of its life.

- **By the time** you are two your head is almost the same size as it will be when you are fully grown.

- **Your brain** will be nearly fully grown by the time you are six.

- **The end of long bones**, such as the arm and leg bones are made of cartilage in children. As we grow this cartilage gradually grows to make the bones longer.

▶ *Your final height depends largely on the height of your parents, but a good diet and a healthy childhood will also affect how tall you will be.*

3 years old

- **By the time** we are fully grown the cartilage has been replaced by bone.

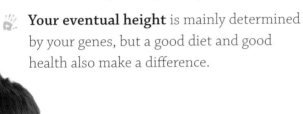

- **Baby boys** grow faster than baby girls during the first seven months.

- **After this**, boys and girls grow at roughly the same rate until they reach puberty.

- **Your eventual height** is mainly determined by your genes, but a good diet and good health also make a difference.

10 years old

6 years old

289

Childhood milestones

DID YOU KNOW?

You learn faster in the first few years of your life than at any other time.

As we grow we learn basic skills such as walking and talking. These are called developmental milestones.

Babies usually start to smile at around four to six weeks.

By six to ten months, we are crawling and pulling ourselves upright.

Toddlers can start walking at any time from nine to 18 months. By the age of two they will be able to throw and kick a ball.

▶ Toddlers start to grasp objects at about eight months and will be able to build a tower from around 18 months onwards.

▶ *Children usually start to draw between the ages of one and two years old.*

- **Although babies** are born with a grasping reflex, it takes at least six months to learn how to pick something up just using our fingers and thumbs.

- **At 12 months** onwards we start to scribble, and by the age of three we can draw a straight line.

- **We usually say** our first word at around 12–14 months old and start to put words together between 14 months and two years.

- **We can talk** in sentences by three-and-a-half years old and learn to read at between five and six years old.

- **Between eight and 12 months** we learn to eat with our fingers and between 18 months and two years we learn to eat with a knife and fork.

- **By the time we are five years old** we have learnt all the basic skills we need in life, and can talk, walk, run, play, eat, dress and make friends.

291

Puberty in boys

Puberty is the time of life when boys mature sexually. It is also a time of rapid growth.

The age of puberty varies, but on average it is between 11 and 15 years for boys. The exact age depends on your genes, general health and weight.

Puberty is started by two hormones produced by the pituitary gland – follicle stimulating hormone and luteinizing hormone.

These hormones encourage the testes to produce the sex hormone testosterone.

Testosterone causes physical changes and a desire for sexual intercourse.

During puberty in a boy, the testes grow and hair sprouts on his face, under his arms and around his genitals. Hair may also grow on the chest.

◄ The voice deepens at puberty in boys. This is sometimes erratic, causing the voice to 'break' and make odd noises.

292

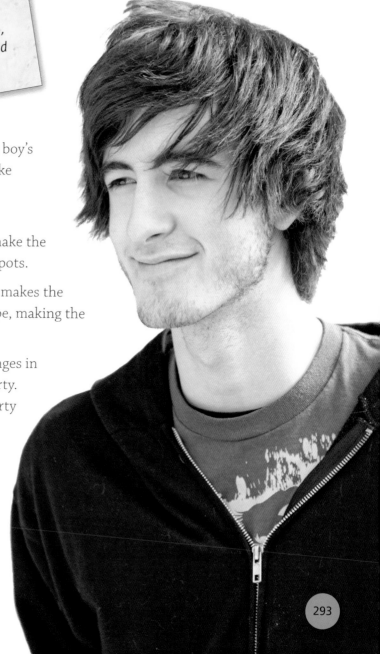

- **Inside his body**, a boy's testes begin to make sperm.

- **Higher levels** of testosterone can make the skin oily, causing spots.

- **Testosterone** also makes the larynx change shape, making the voice deeper.

- **The body** also changes in shape during puberty. By the end of puberty most males have heavier bones and twice as much muscle.

▶ *Facial hair is one of the signs of puberty in boys.*

293

Puberty in girls

- **At puberty**, girls mature sexually and grow suddenly.

- **The age of puberty** varies hugely depending on your genes, weight and general health, but on average it is between ten and 13 years for girls.

- **Two hormones** produced by the pituitary gland – follicle-stimulating hormone and luteinizing hormone – start puberty.

- **These hormones stimulate** the ovaries to produce the sex hormones oestrogen and progesterone.

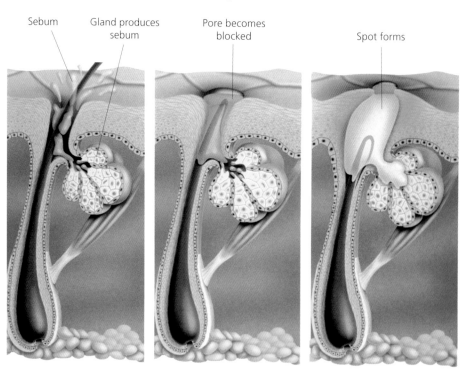

▲ During puberty, glands in the skin produce extra sebum, a type of oil, that can block pores, causing spots.

▲ *Washing your face regularly with a cleanser and warm water will help to prevent spots.*

Oestrogen and progesterone spur the development of a girl's sexual organs and control her monthly menstrual cycle.

During puberty, a girl will develop breasts and grow hair under her arms and around her genitals.

The uterus and ovaries will become bigger.

Increased levels of sex hormones also make the skin oily, causing spots.

A girl's body also changes shape during puberty. The hips widen and body fat increases, especially around the hips, buttocks and thighs.

A year or so after puberty begins, a girl has her menarche (the first menstrual period). When her periods come regularly, she will be able to have a baby.

295

Ageing

Most people live for between 60 and 100 years, although a few live even longer than this.

The longest officially confirmed age is that of a French woman Jeanne Calment, who died in 1997, aged 122 years and 164 days.

Life expectancy is how long statistics suggest you are likely to live.

On average in Europe, men can expect to live about 75 years and women about 80. However, because health is improving generally, people are now living longer.

As adults grow older, their bodies begin to deteriorate (fail). Senses such as hearing, sight and taste weaken.

Hair goes grey as pigment (colour) cells stop working.

Muscles weaken as fibres die.

Bones become more brittle as they lose calcium. Cartilage shrinks between joints, causing stiffness.

◀ *People in Japan have a long life expectancy. This is probably due to a combination of factors including a healthy diet and certain social customs, which tend to favour the elderly.*

Skin wrinkles as the rubbery elastin and collagen fibres that support it sag. Exposure to sunlight speeds this up, which is why the face and hands get wrinkles first.

Circulation and breathing weaken. Blood vessels may become stiff and clogged, forcing the heart to work harder and raising blood pressure.

DID YOU KNOW?
In the UK, there are more people aged over 60 than there are aged under 16.

▼ *Changes in health care mean more and more people than ever before are keeping fit and healthy into old age.*

Metabolism

Body cells are constantly repairing themselves or creating new cells, or substances are constantly being broken down or built up into new substances.

Thousands of these chemical reactions take place in the body every day to keep us alive and healthy. This is metabolism.

There are two types of metabolism – anabolism and catabolism.

In anabolism, complex substances are made from simple ones to build or repair cells. In catabolism, simple substances are broken down to provide energy.

We get energy from food and this energy is used by the cells in the body to keep us healthy and alive.

The basal metabolic rate (BMR) is the amount of energy you need to keep your body going.

Even when you are asleep or doing nothing your body needs energy to keep working. Your heart still beats, you still need to breathe and your temperature needs to be kept normal.

Your BMR decreases with age as you need less energy to keep going as you get older. Exercise increases BMR.

Girls usually have a lower BMR than boys, but women who are pregnant or breastfeeding need more energy and so will have higher BMRs than other women.

◄ *People with a physical job, such as fire fighters, need more energy than people who work at a desk all day.*

299

The hypothalamus

- **The hypothalamus** is a tiny area of the brain that controls the glands in the body and metabolism.

- **It helps** to regulate sleep, hunger, thirst, the body's temperature, blood pressure and fluid balance.

- **Nerve signals** are constantly being sent to and from the hypothalamus to make sure our body is at the right temperature, and our blood pressure and heart rate are correct.

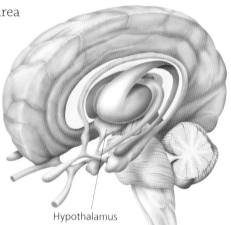

Hypothalamus

▲ *The hypothalamus is a complex area of the brain with a number of important functions.*

- **The hypothalamus** also releases hormones called releasing hormones that act on the pituitary gland.

- **These hormones** encourage the pituitary gland to release hormones that affect growth and reproduction.

- **The hypothalamus** controls the circadian cycle. This is our internal body clock.

DID YOU KNOW?

The word circadian comes from the Latin words 'circa', meaning around, and 'diem', meaning day.

- **Your body clock** is set to a 24-hour cycle.

- **The cycle** is connected to daylight and tells us when we should be alert, eating and sleeping.

- **If you fly around the world** to a very different time zone you will get jet lag. Your internal body clock is still set on the time at home and your body rhythms are upset.

- **The hypothalamus** also controls hunger and thirst.

▼ *When we travel across time zones we get 'jet lag' because our internal body clock needs a few days to adjust to the different time zone.*

Temperature

The inside of your body stays at a constant temperature of around 37°C (98°F), rising a few degrees only when you are ill.

Your body creates heat by burning food in its cells, especially the 'energy sugar' glucose.

Even when you are resting, your body generates so much heat that you are comfortable only when the air is slightly cooler than you are.

When you are working hard, your muscles can generate as much heat as a 2 kW heater (a typical room heater).

▶ The body's temperature can be easily monitored using a digital thermometer.

Your body loses heat as you breathe in cool air and breathe out warm air. Your body also loses heat by giving it off from your skin.

The body's temperature control is the tiny hypothalamus in the lower front of the brain.

Temperature sensors in the skin, in the body's core, and in the blood tell the hypothalamus how hot or cold your body is.

- **If it is too hot**, the hypothalamus sends signals to your skin telling it to sweat more. Signals also tell blood vessels in the skin to widen – this increases the blood flow, increasing the heat loss from your blood.

DID YOU KNOW?

Your body temperature changes during the day – it will be about half a degree lower in the middle of the night than in the afternoon.

- **If it is too cold**, the hypothalamus sends signals to the skin to cut back skin blood flow, as well as signals to tell the muscles to generate heat by shivering.

- **If it is too cold**, the hypothalamus may also stimulate the thyroid gland to send out hormones to make your cells burn energy faster and so make more heat.

▲ On a hot day, the skin becomes flushed as blood vessels widen to try to lose heat. Drinking something cold can help to cool us down.

Health and disease

- **Health is partly determined** by your genes. Some diseases run in families and you are more likely to get them no matter what you do.

DID YOU KNOW?

To keep healthy, you need to exercise for at least 30 minutes every day.

- **Other diseases** are caused by an unhealthy lifestyle. If you eat well, exercise and look after yourself you are more likely to live a long and healthy life.

- **You should eat** a good balance of foods. Try to eat plenty of fruit and vegetables every day and only eat fatty food occasionally.

- **Regular exercise** is important to keep your bones strong and your muscles healthy.

- **Exercise also helps** to keep your heart healthy and keeps you flexible.

- **If you want** to keep healthy you should not smoke cigarettes, as the tobacco in cigarettes damages lungs and blood vessels as well as other parts of the body.

- **You should also** not drink too much alcohol. Too much alcohol damages the liver and brain.

- **Illness occurs** when your body does not function normally. An illness may be caused by an injury, infection or disease.

- **Regular checkups** by your doctor will spot signs of disease early.

- **You should also** make sure you have any vaccinations that your GPs suggests.

TOP 10 CAUSES OF DEATH	CONTRIBUTING FACTORS
HEART ATTACK	An unhealthy diet and lack of exercise increase the risk of a heart attack
STROKE	A fatty diet and stress increase the risk of brain damage because the blood supply is interrupted
CHEST INFECTION	In undeveloped countries chest infections can be fatal because of a lack of medicine
LUNG DISEASE	Breathing problems caused by smoking cigarettes can be avoided by not smoking
DIARRHOEA	Infected water and food is common in poor countries. Gut infections can kill without treatment
HIV/AIDS	The majority of HIV infections are acquired through unprotected sex
TUBERCULOSIS	Developed countries can prevent tuberculosis with vaccines, but these are rare in undeveloped countries
LUNG CANCER	This is usually caused by smoking cigarettes
ROAD TRAFFIC ACCIDENTS	Poor driving is a killer in many countries
PREMATURE BIRTHS	Many countries do not have good enough health care for babies that are born too early

Diet

- **Your diet** is what you eat. A good diet includes the correct amount of proteins, carbohydrates, fats, vitamins, minerals, fibre and water.

- **Most of the food** you eat is fuel for the body, provided mostly by carbohydrates and fats.

- **Carbohydrates are foods** made from kinds of sugar, such as glucose and starch. These are found in foods such as bread, rice, potatoes and sweet things.

- **Fats are greasy foods** that will not dissolve in water. Some, such as the fats in meat and cheese, are solid. Some, such as cooking oil, are liquid.

- **Fats are not usually burned up** straight away, but are stored around your body until they are needed.

- **Proteins are needed** to build and repair cells. They are made from special chemicals called amino acids.

- **Fibre or roughage** is supplied by cellulose from plant cell walls. Your body cannot digest fibre, but needs it to keep the bowel muscles exercised.

- **Many ready-made foods** now come with food labels on them so that you can see how much carbohydrate, fat and protein they contain.

- **People in different countries** have different sorts of diets. Japanese people eat more fish and rice than people in the UK.

◀ *It is important to eat a healthy, balanced diet with plenty of fruit and vegetables, fish and lean meat, fibre and some fat.*

The Mediterranean diet, which is eaten by people in areas such as Italy and Greece, contains plenty of fruit, vegetables, fish and olive oil and is particularly healthy.

309

Carbohydrates

- **Your body's main source** of energy is carbohydrates in food. They are plentiful in sweet things and in starchy food such as bread, cakes and potatoes.

- **Carbohydrates are burned** by the body in order to keep it warm and to provide energy for growth and muscle movement, as well as to maintain basic body processes.

- **Carbohydrates are among the most common** of organic (life) substances – plants, for instance, make carbohydrates by taking energy from sunlight.

- **Chemical substances** called sugars are also carbohydrates. Sucrose (the sugar in sugar lumps and caster sugar) is just one of these sugars.

- **Simple carbohydrates** such as glucose, fructose (the sweetness in fruit) and sucrose are sweet and soluble (they will dissolve in water).

- **Complex carbohydrates** (or polysaccharides) such as starch are made when molecules of simple carbohydrates join together.

- **A third type** of carbohydrate is cellulose.

- **The carbohydrates you eat** are turned into glucose for your body to use at once, or stored in the liver as the complex sugar glycogen (body starch).

- **The average adult** needs 2000–3000 Calories a day.

- **A Calorie** is the heat needed to warm one litre of water by 1°C.

▶ Bread and other wheat products, rice and potatoes are all full of starch, a complex carbohydrate that gives us a steady supply of energy.

Glucose

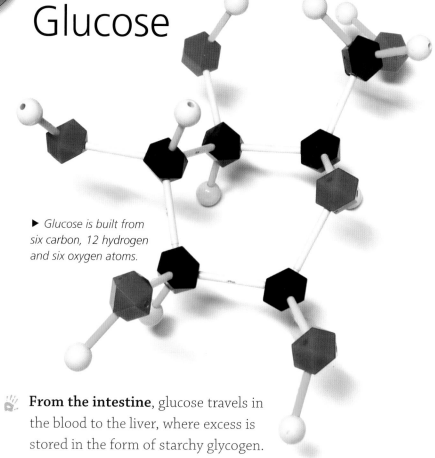

▶ Glucose is built from six carbon, 12 hydrogen and six oxygen atoms.

- **From the intestine**, glucose travels in the blood to the liver, where excess is stored in the form of starchy glycogen.

- **For the body to work effectively**, levels of glucose in the blood (called blood sugar) must always be correct.

- **Glucose is the body's energy chemical**, used as the fuel in all cell activity.

- **Glucose is a kind of sugar** made by plants as they take energy from sunlight. It is commonly found in many fruits and fruit juices, along with fructose.

- **The body gets its glucose** from carbohydrates in food, broken down in stages in the intestine.

- **Blood sugar levels** are controlled by two hormones, glucagon and insulin, sent out by the pancreas.

- **When blood sugar is low**, the pancreas sends out glucagon and this makes the liver change more glycogen to glucose.

- **When blood sugar is high**, the pancreas sends out insulin and this makes the liver store more glucose as glycogen.

- **Inside cells**, glucose may be burned for energy, stored as glycogen, or used to make triglyceride fats.

▼ *A blood glucose monitor checks that a person's blood sugar is at a healthy level.*

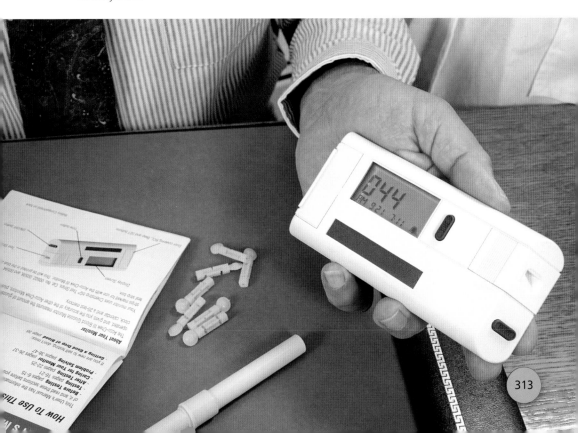

Proteins

- **Proteins are needed** for growth and repair.

- **They are made** from simple substances called amino acids.

- **Proteins are broken down** in the body into amino acids. These are then used to make different proteins within the body.

- **There are 20** different amino acids. Your body can make 11 of them. The other nine are called essential acids and they come from food.

DID YOU KNOW?

Most plants can make 20 amino acids, but humans can only make 11.

◄ Meat, fish and dairy products such as eggs and cheese are our most common sources of protein.

▶ Vegetarians do not eat meat or fish, so must be careful to eat plenty of other sources of protein, such as chickpeas.

Foods that are high in protein include meat, fish, eggs, milk, cheese, nuts, lentils and beans.

A correctly balanced vegetarian diet can provide all the essential amino acids.

About one-sixth of a healthy diet should consist of protein.

People who have an allergy to wheat are allergic to gluten, the protein in wheat.

Some people are allergic to the proteins in peanuts or seafood, or to casein, the protein in milk.

◀ Nuts, such as these almonds, are good sources of protein and contain vitamins and minerals too.

315

Fats

▼ *Fat cells are numerous under the skin, providing your body with a store of energy and a layer of insulation to keep you warm.*

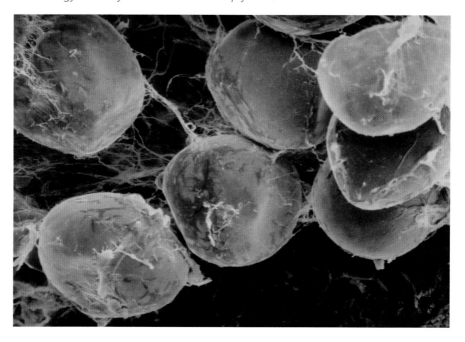

Fats are an important source of energy. Together with carbohydrates and proteins, they make up your body's three main components of foods.

While carbohydrates are generally used for energy immediately, your body often stores fat to use for energy in times of shortage.

Weight for weight, fats contain twice as much energy as carbohydrates.

Fats are important organic (life) substances, found in almost every living thing. They are made from substances called fatty acids and glycerol.

- **Food fats** are greasy vegetable or animal fats that will not dissolve in water.

- **Most vegetable fats** such as corn oil and olive oil are liquid, although some nut fats are solid.

- **Most animal fats**, as in meat, milk and cheese, are solid. Milk is mainly water with some solid animal fats. Most solid fats melt when warmed.

- **Fats called triglycerides** are stored around the body as adipose tissue (body fat). These act as energy stores and also insulate the body against the cold.

- **Fats called phospholipids** are used to build body cells.

- **In your stomach**, bile from your liver and enzymes from your pancreas break fats down into fatty acids and glycerol. These are absorbed into your body's lymphatic system or enter the blood.

▶ *Cheese contains saturated fat, which is linked to high levels of the substance cholesterol in the blood and may increase the risk of a heart attack.*

Vitamins

- **Vitamins are special substances** the body needs to help maintain chemical processes inside cells.

- **There are at least** 15 known vitamins. A lack of any vitamin in the diet can cause certain illnesses.

- **The first vitamins** discovered were given letter names such as B. Later discoveries were given chemical names, such as E vitamins, which are known as tocopherols.

- **Some vitamins** such as A, D, E and K dissolve in fat and are found in animal fats and vegetable oils. They may be stored in the body for months.

- **Vitamin C** is found in fruit such as oranges, tomatoes and fresh green vegetables.

▶ Citrus fruit, such as oranges, lemons and limes, and green vegetables are full of vitamins, which is why they are so important in our diet.

- **Before the 18th century**, sailors on long voyages used to suffer from the disease scurvy, caused by a lack of vitamin C in their diet.

- **There are several** B vitamins. These dissolve in water and are found in green leaves, fruits and cereal grains. Many are also in meat and fish. They are used daily.

- **Vitamins D and K** are the only ones made inside the body. Vitamin K is made by bacteria in the gut.

- **Vitamin D** is made in the skin when we are out in the sun; it is essential for bone growth in children.

Minerals

- **Minerals are substances** needed by your body to keep it healthy.

- **You need** about 20 different minerals. Most minerals are simple chemicals, such as calcium.

- **You only need** tiny amounts of each mineral every day but without them you will become ill.

- **Your body** can store some minerals but needs a regular supply of others in your food.

▶ *Many toothpastes have the mineral fluoride added to them to help strengthen teeth and prevent tooth decay.*

◄ Spinach is a good source of iron, which may explain why the cartoon character Popeye the sailor eats it to give himself super strength!

- **A healthy balanced diet** that includes plenty of fruit and vegetables should supply all the minerals that your body needs.

- **Iron can be found** in meat, eggs, and leafy green vegetables. You need iron to help make red blood cells.

- **Calcium helps** build strong bones and teeth. It is found in dairy products such as milk and cheese and fish with bones, such as sardines.

- **Healthy bones** also need magnesium, which is found in nuts, seafood and cocoa.

- **One of the hormones** produced by the thyroid gland contains iodine, which is found in red meat, nuts and leafy green vegetables.

- **You also need minerals** such as selenium for normal heart and liver function, and zinc for growth and energy.

321

Water

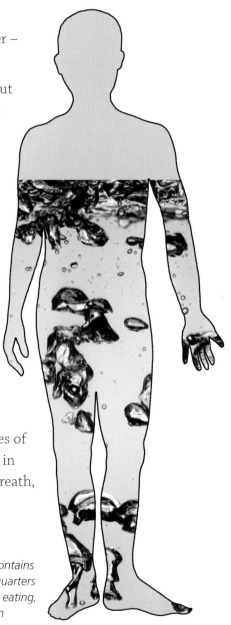

- **Your body** is mainly made of water – more than 60 percent.

- **You can survive for weeks** without food, but no more than a few days without water.

- **You gain water** by drinking and eating, and as a by-product of cell activity.

- **You lose water** by sweating and breathing, and in your urine and faeces.

- **The average person takes** in 2.5 litres of water a day – 1.4 litres in drink and 0.8 litres in food. Body cells add 0.3 litres, bringing the total water intake to 2.5 litres.

- **The average person loses** 1.5 litres of water every day in urine, 0.5 litres in sweat, 0.3 litres as vapour in the breath, and 0.2 litres in faeces.

▶ *Your body is mostly water. Even bone contains one-fifth water, while your brain is three-quarters water. You take it in through drinking and eating, and lose it by urinating, sweating and even breathing.*

The **water balance** in the body is controlled mainly by the kidneys and adrenal glands.

The **amount of water** the kidneys let out as urine depends on the amount of salt there is in the blood.

If you drink a lot, the saltiness of the blood is diluted (watered down). To restore the balance, the kidneys let out a lot of water as urine.

▶ If you sweat a lot during heavy exercise, you need to make up for all the water you have lost by drinking. Your kidneys make sure that if you drink too much, you lose water as urine.

If you drink little or sweat a lot, the blood becomes more salty, so the kidneys restore the balance by holding on to more water.

323

Body salts

Body salts are not simply sodium chloride – the salt some people sprinkle on food – they are an important group of chemicals that play a vital role in your body.

Examples of components in body salts include potassium, sodium, manganese, chloride, carbonate and phosphate.

Body salts are important in maintaining the balance of water in the body, and on the inside and the outside of body cells.

The body's thirst centre is the hypothalamus. It monitors salt levels in the blood and sends signals telling the kidneys to keep water or to let it go.

You gain salt from the food you eat.

You can lose salt if you sweat heavily. This can make muscles cramp, which is why people take salt tablets in the desert or drink a weak salt solution.

Too much salt in food may result in high blood pressure in certain people.

When dissolved in water, the chemical elements that salt is made from split into ions – atoms or groups of atoms with either a positive or a negative electrical charge.

The balance of water and salt inside and outside of body cells often depends on a balance of potassium ions entering the cell and sodium ions leaving it.

▼ *Athletes often drink special sports drinks to make up for the loss of salt caused by sweating.*

Fibre

- **Fibre is material in food** that the body cannot break down.

- **Partly digested food** in the gut is given bulk by fibre, making it easier for the muscles in the gut to move the food along.

- **Fibre also** softens partly digested food, making it easier to excrete.

- **Fibre generally** passes out of the body undigested.

- **There are two types** of fibre – soluble fibre, which dissolves in water, and insoluble fibre, which does not dissolve.

- **Soluble fibre** is found in oats, some fruits and vegetables and beans.

▼ *Beans, bananas, seeds and nuts are all good sources of fibre.*

▶ *Peanuts contain a surprising amount of fibre, as well as healthy fats and protein.*

- **Insoluble fibre** is found in whole grains, wheat, nuts, seeds and some vegetables.

- **Foods that contain fibre** often help to fill you up so that you are less hungry.

- **People who eat foods** containing plenty of fibre are less likely to get some diseases of the large intestine, including cancer.

- **Some types** of fibre, especially soluble fibre, lower levels of fat in the blood and help to prevent heart disease.

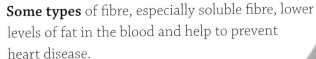

Exercise

Exercise is any activity that keeps you fit and healthy. Some people do exercise by playing a sport or going to the gym. Others take walks or do something around the house such as gardening.

People exercise because they enjoy it, to keep fit or to lose weight.

When you exercise, your muscles have to work much harder than normal, so need much more oxygen and glucose (a kind of sugar) from the blood.

To boost oxygen, your heart beats twice as fast and pumps twice as much blood, and your lungs take in ten times more air with each breath.

To boost glucose, adrenalin triggers your liver to release its store of glucose.

If oxygen delivery to muscles lags, the muscles fill up with lactic acid, affecting your body for hours afterwards and sometimes causing painful cramp.

◄ Yoga is an excellent exercise for building up flexibility, strength and balance without putting too much stress on the body.

▲ *Walking is great exercise as well as a good way to spend time with family and friends.*

The fitter you are, the quicker your body returns to normal after exercise.

It is always important to wear the right equipment, such as football boots for playing football, and to start slowly to get your muscles warmed up.

Regular exercise multiplies muscle fibres and strengthens tendons. It can help reduce weight when it is combined with a controlled diet.

329

Types of exercise

Regular exercise is important to improve your strength, endurance (stamina or staying power) and flexibility (suppleness).

Strength is how strong your muscles are. A weightlifter is very strong.

Stamina helps you to exercise for a long time without becoming tired. A marathon runner has lots of stamina.

Suppleness is how flexible you are. A gymnast is very supple.

Different types of exercise help to develop strength, stamina and suppleness. Some exercises are good for developing one and some are good for developing all three.

Walking and running are good for developing stamina.

Sports such as football and swimming are good for developing strength, stamina and suppleness.

◀ Cycling is one of the best all round exercises there is – it builds up stamina and strength as well as being good for suppleness and balance.

▲ Team sports such as basketball are good aerobic exercises and great fun too.

DID YOU KNOW?
When you exercise hard, your body burns up energy 20 times as fast as normal.

- **Aerobic exercise** is exercise that is long and hard enough for the oxygen supply to the muscles to rise enough to match the rapid burning of glucose.

- **Regular aerobic exercise** strengthens your heart and builds up your body's ability to supply extra oxygen through your lungs to your muscles.

- **In anaerobic exercise**, muscles are used hard for short periods of time. The oxygen supply to the muscles cannot keep up and they have to use other forms of energy.

331

Fitness

- **The word fitness refers** to how much and what kind of physical activity you can do without getting tired or strained.

- **Fitness depends** on your strength, flexibility (suppleness) and endurance (stamina).

- **One key to fitness** is cardiovascular fitness – that is, how well your heart and lungs respond to the extra demands of exercise.

- **One measure** of cardiovascular fitness is how quickly your pulse rate returns to normal after exercise – the fitter you are, the quicker it returns.

◄ *Skiing is one of the most demanding of all sports, and top skiers need to be extremely fit to cope with the extra strain on their bodies.*

▲ *Many people keep fit by attending exercise classes.*

DID YOU KNOW?
Children and teenagers should get at least one hour of exercise every day.

Another measure of cardiovascular fitness is how slowly your heart beats during exercise – the fitter you are, the slower it beats.

Being fit improves your physical performance. It can often protect against illness and even slow down the effects of ageing.

Cardiovascular fitness reduces the chances of getting heart disease.

Fitness tests involve comparing such things as height, weight and body fat, and measuring blood pressure and pulse rate before and after exercise.

333

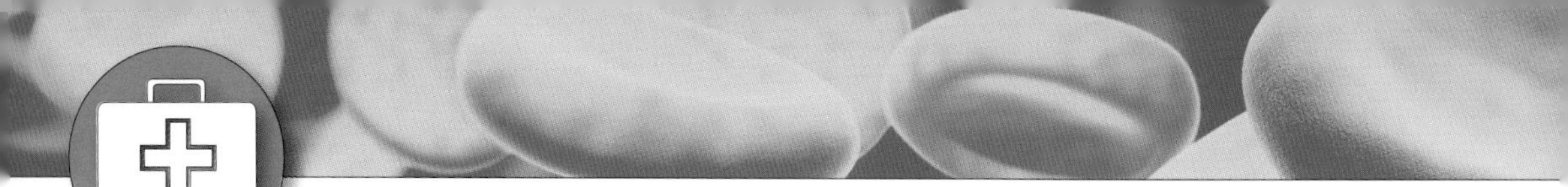

Types of disease

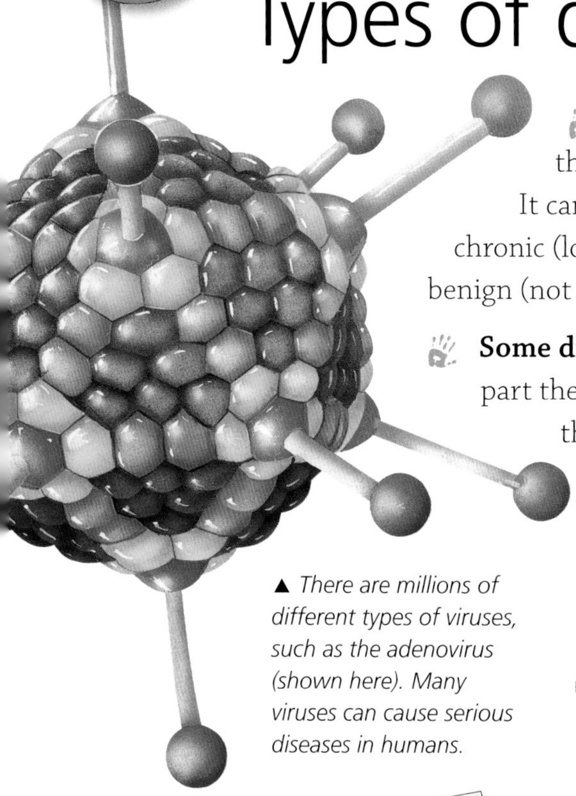

A **disease** is something that upsets the normal working of any living thing. It can be acute (sudden, but short-lived), chronic (long-lasting), malignant (spreading) or benign (not spreading).

Some diseases are classified by the body part they affect (such as heart disease), or by the body activity they affect (such as respiratory, or breathing, disease).

Heart disease is the most common cause of death in the USA, Europe and also Australia.

▲ *There are millions of different types of viruses, such as the adenovirus (shown here). Many viruses can cause serious diseases in humans.*

Some diseases are classified by their cause. These include the diseases caused by the staphylococcus bacteria – pneumonia is one such disease.

Contagious diseases are caused by germs such as bacteria and viruses. They include the common cold, polio, flu and measles. Their spread can be controlled by good sanitation and hygiene, and also by vaccination programmes.

DID YOU KNOW?

The most common disease in the world is tooth decay – so remember to brush your teeth!

Non-contagious diseases may be inherited or they may be caused by such things as eating harmful substances, poor nutrition or hygiene, or being injured.

- **Non-contagious diseases** may also be caused by body cells acting wrongly and attacking the body's own tissues. This type of disease is called an autoimmune disease.

- **Degenerative diseases** occur in older people as the body's tissues start to get older and either do not function normally or gradually disappear.

- **Endemic diseases** are diseases that occur in a particular area of the world, such as sleeping sickness in Africa.

- **Diseases can be** either contagious (passed on by contact) or non-contagious.

▼ *This microscopic picture shows a cancer cell.*

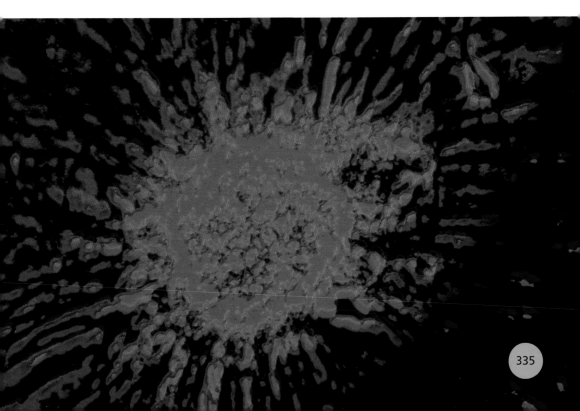

335

Defence

- **The human body** has several defences to prevent disease affecting it.

- **Our first defence** is the skin. The skin acts as a physical barrier to germs and helps to prevent injury.

- **The skin also produces** sweat and oils that help to keep the skin moisturized and prevent germs from growing.

- **Tears keep our eyes moist** and help wash away dust and dirt. They also contain an antibacterial agent to help destroy germs.

- **Saliva in the mouth** helps to kill bacteria that cause tooth decay and gum disease.

- **Mucus in the lining of the airways** helps to trap dust and germs.

- **Stomach acid** is strong enough to kill most germs that may be eaten or taken in with drink.

- **Good bacteria** in the gut prevent harmful bacteria from growing.

- **Helpful bacteria** that live in a woman's vagina help to keep it slightly acidic. This stops harmful bacteria from growing.

- **The main defence** against disease once germs have entered the body is the body's immune system.

▶ The gut normally contains good bacteria (shown here in pink) that help to boost the immune system and prevent the growth of disease-causing bacteria.

336

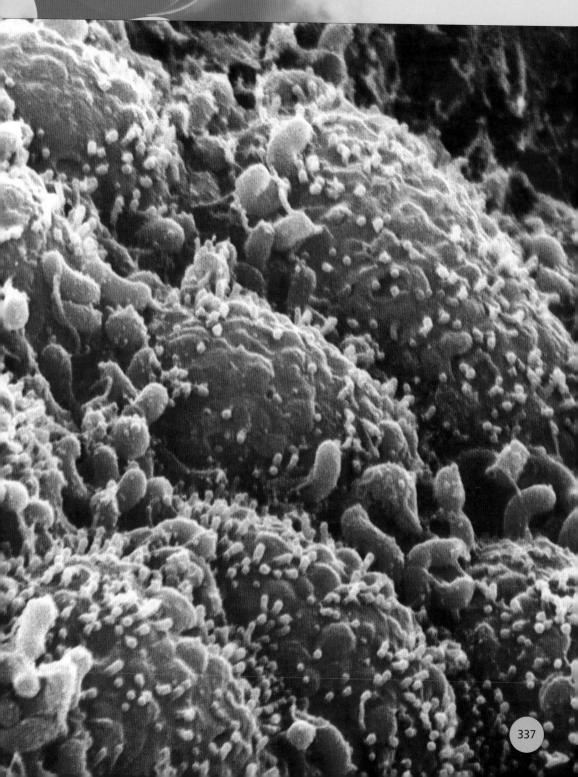

Germs

- **Germs are microscopic organisms** that enter your body and cause harm.

- **The scientific word** for germ is 'pathogen'.

- **When germs** begin to multiply inside your body, you are suffering from an infectious disease.

- **An infection that spreads** throughout your body (flu or measles, for example) is called a systemic infection.

- **An infection that affects** only a small area (such as dirt in a cut) is called a localized infection.

▼ *Once germs such as Anthrax (blue) or E. coli (purple) bacteria enter the body, they multiply rapidly, causing disease and making us feel unwell.*

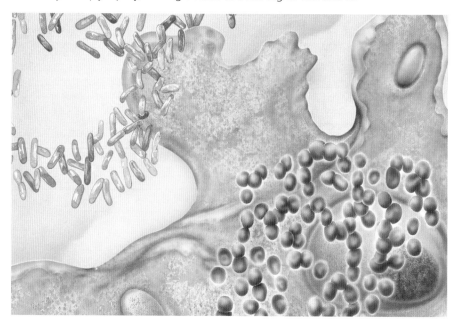

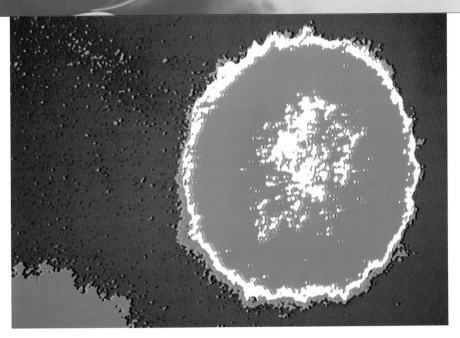

▲ *The disease AIDS (Acquired Immune Deficiency Syndrome) is caused by a virus called HIV (Human Immunodeficiency Virus). This virus gets inside vital cells of the body's immune system and weakens its ability to fight against other infections.*

It is often the reaction of your body's immune system to the germ that makes you feel ill.

There are several types of germ, including bacteria, viruses, worms and fungi.

Germs may enter the body through an injury, such as a cut or a scrape.

Many germs, such as the ones that cause colds, are breathed in.

Some germs can be spread just by being near someone who has an infection. Others can only cause an infection if blood from one person mixes with another.

DID YOU KNOW?
Germs vary in size from microscopic viruses to worms that can grow to several metres long.

339

Bacteria

- **Bacteria** are single-celled organisms. They are found almost everywhere in huge numbers, and multiply rapidly.

- **There are thousands** of different types of bacteria but most are harmless. Some even do us good.

DID YOU KNOW?

Bacteria can live anywhere, even in regions where there are no living animals or plants.

- **Bacteria can be divided** into three groups – cocci are round cells, spirilla are coil-shaped, and bacilli are rod-shaped.

- **Bacteria usually cause disease** by producing harmful chemicals called toxins. The toxins enter cells in the body and destroy them.

- **Some types of bacteria** do not produce toxins but enter body cells instead.

- **Once the bacteria** are in the cells they multiply until they burst the cell. All the new bacteria then find new cells to enter.

- **Antibiotics** are used to treat bacterial infections. These kill the bacteria in the body or stop then multiplying so that the body has a chance to destroy them.

- **Bacteria commonly cause infections** that affect the airways and lungs. Tuberculosis is caused by bacteria.

- **Bacteria also cause diseases** such as tetanus and typhoid. Bacteria that enter the blood stream can cause blood poisoning.

- **The 'Black Death'** that killed millions of people in Europe in the 1340s was actually a bacterial infection called the bubonic plague.

▼ *These rod-shaped bacteria have hairs so they can stick to other cells, and tail-like projections to help them move.*

Viruses

▼ *Once a virus has entered a body cell, it replicates and then leaves the cell to infect more cells.*

- **Millions of viruses** can fit inside a single cell. They are the smallest living organisms in the world.

- **Viruses can only live** and multiply by taking over other cells – they cannot survive on their own.

- **They consist of** genetic material surrounded by a protective coat.

- **Viruses cause disease** by entering body cells. Once inside a cell they reproduce and leave the cell to go on and infect more cells.

- **Common diseases** such as colds, flu, mumps, measles and chickenpox are all caused by viruses.

- **Viruses also cause** severe infections such as AIDS and fevers associated with bleeding.

- **Viral infections** can be prevented by vaccinations.

- **An epidemic occurs** if many more people than usual get a viral infection. A pandemic occurs if the epidemic affects people around the world.

- **In 1918** the flu virus infected about a third of the world's population and it is thought that about 50 million people died.

DID YOU KNOW?

There are over 200 types of virus that can cause a common cold.

- **Doctors worry** that this might happen again with a different type of flu virus.

Fungi, protozoa and parasites

- **Parasites are animals** such as worms that may live in or on your body, feeding on it and making you ill.

- **There are two types** of worms that can infect humans – roundworms and flatworms.

- **Worms usually** infect people who eat contaminated food. They usually live in the gut.

- **Tapeworms** can grow to over 10 m long in the gut.

- **Fungi** and tiny organisms called protozoa can also cause illness.

- **Fungi form spores** that may cause disease if you breathe them in. These spores can stay in your body for many years before you become ill.

- **Protozoal infections** are often transmitted by eating or drinking food or water that has been infected, or by insect bites.

- **Malaria** is a serious disease caused by protozoa.

- **Food poisoning** in tropical countries is often caused by infection with a protozoa.

- **Most infections** with fungi, protozoa and worms can be easily treated with drugs.

▶ *Worms can be either flat (flatworms) or round (roundworms). Tapeworms can grow to several metres in length, but pinworms are tiny.*

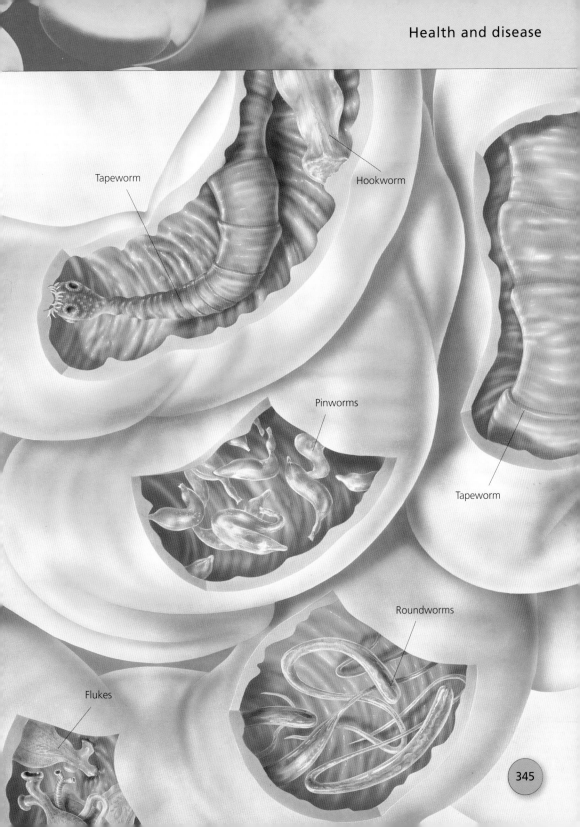

Tapeworm

Hookworm

Pinworms

Tapeworm

Roundworms

Flukes

Diagnosis

Diagnosis is when a doctor works out what a patient is suffering from – the illness and perhaps its cause.

The history is the patient's own account of their illness. This provides the doctor with a lot of clues.

The **prognosis** is the doctor's assessment of how the illness will develop in the future.

Symptoms are changes that the patient or others notice and report.

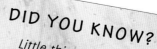

DID YOU KNOW?
Little things, like the colour and shape of your nails, can help your doctor make a diagnosis.

Signs are changes the doctor detects on examination and maybe after tests.

After taking a history the doctor may carry out a physical examination, looking at the patient's body for symptoms such as swelling and tenderness.

A stethoscope is a set of ear tubes that allows the doctor to listen to body sounds, such as breathing and the heart beating.

With certain symptoms, a doctor may order laboratory tests of blood and urine samples. Devices such as ultrasounds and X-rays may also be used to take special pictures.

Doctors nowadays may use computers to help them make a diagnosis.

◀ When a doctor examines a patient, she is looking for clues that will help her to make a diagnosis.

347

Tests

- **A doctor** may order tests to help him or her find out why you are not feeling well.

- **Tests may also** be done regularly to make sure you are healthy, even if you do not feel ill.

- **Tests may look** at types of cells or levels of chemicals.

- **Blood is usually taken** from a blood vessel in the crook of the elbow and may be tested to look at blood cells or to measure chemicals in the blood.

- **Blood tests** are a good way of seeing if there is anything wrong with your body. For example, they can tell whether you have an infection, if your liver, kidneys, immune system and glands are working normally, or whether your sugar levels are normal.

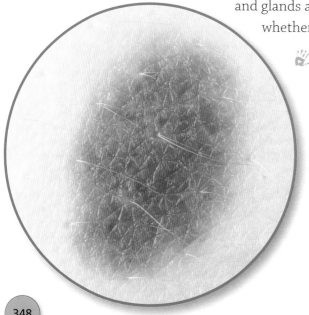

- **Urine may also** be tested for infection, sugar levels and to see if your kidneys are working normally.

◀ A mole that changes colour or shape may need to be removed to test for a form of skin cancer.

348

▲ *Blood pressure measurements will give this doctor an idea of whether his patient is likely to develop heart or circulation problems.*

A sample of faeces can be tested for infections in the gut or to see if your gut healthy.

Sometimes a piece of tissue may need to be tested to look for abnormal cells.

Tissue is usually taken during a small operation.

349

Microscopes

- **Optical microscopes** use lenses and light to magnify things (make them look bigger). By combining two or more lenses, they can magnify specimens up to 2000 times and reveal individual blood cells.

- **To magnify** things more, scientists use electron microscopes – microscopes that fire beams of tiny charged particles called electrons.

- **Electrons have wavelengths** 100,000 times smaller than light and so can give huge magnifications.

- **Scanning electron microscopes** (SEMs) are able to magnify things up to 100,000 times.

- **SEMs show** such things as the structures inside body cells.

- **Transmission electron microscopes** (TEMs) magnify even more than SEMs – up to five million times.

- **TEMs** can reveal the individual molecules in a cell.

- **SEM specimens** (things studied) must be coated in a special substance such as gold. They give a three-dimensional view.

- **Optical microscope specimens** are thinly sliced and placed between two glass slides. They give a cross-sectional view.

- **Microscopes** help to identify germs.

▶ *Microscopes have several lenses that can produce different magnifications. The lenses bend the light shining through the object.*

X-rays

🖐 **X-rays are a form** of electromagnetic radiation, as are radio waves, microwaves, visible light and ultraviolet. They all travel as waves, but have different wavelengths.

🖐 **X-ray waves** are much shorter and more energetic than visible light waves. X-rays are invisible because their waves are too short for our eyes to see.

▼ It is important to keep absolutely still while an X-ray is being taken so that a clear image is obtained.

- **X-rays are made** when negatively charged particles called electrons are fired at a heavy plate made of the metal tungsten. The plate bounces back X-rays.

DID YOU KNOW?
Doctors have been using X-rays to help diagnose illness since the end of the 19th century.

- **Even though** they are invisible to our eyes, X-rays register on photographic film.

- **X-rays** are so energetic that they pass through some body tissues like a light through a net curtain.

- **To make** a photograph, X-rays are shone through the body. The X-rays pass through some tissues and turn the film black, but are blocked by others, leaving white shadows on the film.

- **Each kind** of tissue lets X-rays through differently. Bones are dense and contain calcium, so they block X-rays and show up white on film. Skin, fat, muscle and blood let X-rays through and show up black on film.

- **X-ray radiation** is dangerous in high doses, so the beam is encased in lead, and the radiographer who takes the X-ray picture stands behind a screen.

- **X-rays** are very good at showing up bone defects. So if you break a bone, it will probably be X-rayed.

- **Sometimes a substance** called a contrast medium is used to show up parts of the body more clearly. Contrast medium shows up white on the film.

353

Ultrasound

Ultrasound uses sound waves that are too high for us to hear to produce pictures of the inside of the body.

The sound waves travel through the body and bounce off our organs, like sonar from a submarine.

Ultrasound scans use a device that sends ultrasound waves through the body and picks up the echoes as they bounce back off objects.

By measuring the echoes the scanner can tell how deep in the body the object is, what shape it is and whether it is solid or filled with air.

A computer converts all this information into a picture.

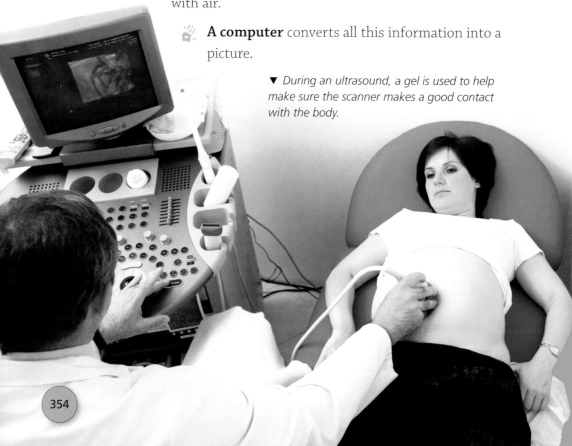

▼ During an ultrasound, a gel is used to help make sure the scanner makes a good contact with the body.

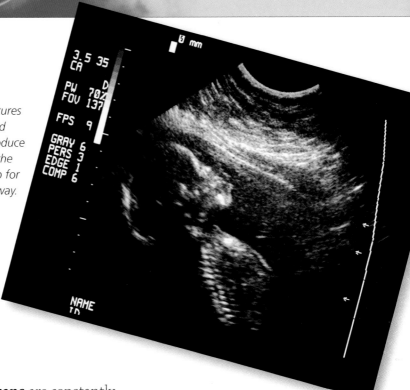

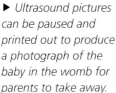

▶ *Ultrasound pictures can be paused and printed out to produce a photograph of the baby in the womb for parents to take away.*

Ultrasound scans are constantly updated through the scanning device and can show movement.

Ultrasound is completely safe and is used to look at a baby in the womb to make sure it is growing normally and is healthy.

DID YOU KNOW?

3D animated ultrasound scans can now be produced, producing a '4D' effect.

Ultrasound is also often used to look at the heart and the heartbeat.

Any of the organs in the body can be examined using ultrasound.

Because ultrasound shows movement it is also often used to look at the blood flow through blood vessels, especially those in the leg and the neck.

355

Scans

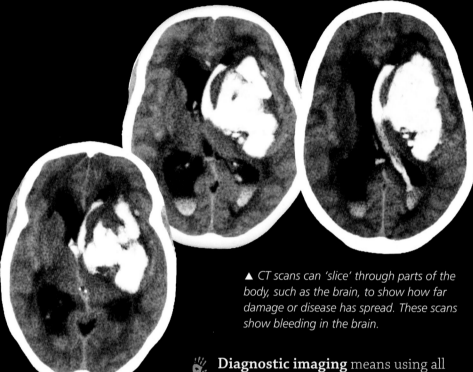

▲ *CT scans can 'slice' through parts of the body, such as the brain, to show how far damage or disease has spread. These scans show bleeding in the brain.*

Diagnostic imaging means using all kinds of complex machinery to make pictures or images of the body to help diagnose and understand a problem.

Many imaging techniques are called scans, because they involve scanning a beam around the patient, to and fro in lines or waves.

CT scans rotate an X-ray beam around the patient while moving him or her slowly forward. This gives a set of pictures showing different slices of the patient's body.

CT stands for computerized tomography.

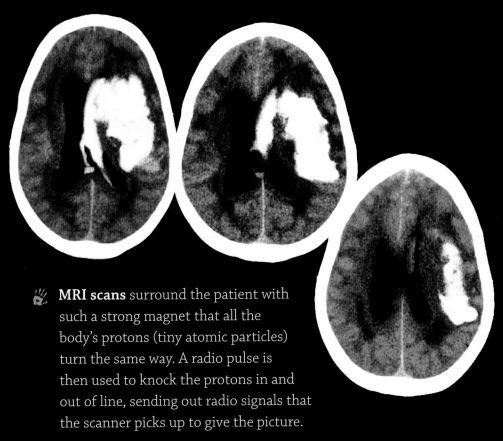

- **MRI scans** surround the patient with such a strong magnet that all the body's protons (tiny atomic particles) turn the same way. A radio pulse is then used to knock the protons in and out of line, sending out radio signals that the scanner picks up to give the picture.

- **MRI stands for** magnetic resonance imaging.

- **PET scans** involve injecting the patient with a mildly radioactive substance, which flows around with the blood and can be detected because it emits (gives out) particles called positrons.

- **PET stands for** positron emission tomography.

- **PET scans** are good for seeing how the brain and heart are functioning.

357

Drugs

- **Antibiotic drugs** are used to treat bacterial infections such as tuberculosis (TB) or tetanus. They were once grown as moulds (fungi) but are now made artificially.

- **Penicillin was the first** antibiotic drug, discovered in a mould in 1928 by Alexander Fleming (1881–1955).

- **Analgesic drugs** such as aspirin relieve pain, working mainly by stopping the body making prostaglandin, the chemical that sends pain signals to the brain.

▼ *Thousands of different drugs are today used to treat illness.*

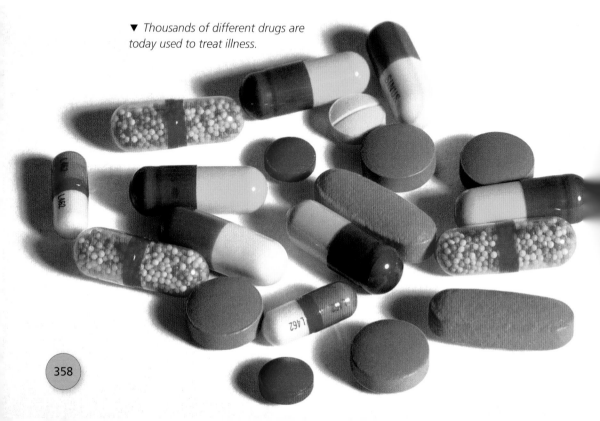

- **Tranquillizers are drugs** that calm. Minor tranquillizers are drugs such as prozac, used to relieve anxiety.

- **Major tranquillizers** are used to treat mental illnesses such as schizophrenia.

- **Psychoactive drugs** change your mood. Many psychoactive drugs, such as heroin, are dangerous and illegal.

- **Stimulants are drugs** that boost the release of the nerve transmitter noradrenaline, making you more lively and awake. They include the caffeine in coffee.

▲ *Alexander Fleming was a British bacteriologist. His discovery in 1928 of the life-saving antibiotic, penicillin, opened a new era for medicine.*

- **Narcotics, such as morphine**, are powerful painkillers that mimic the body's own natural painkiller, endorphin.

- **Depressants are drugs** such as alcohol, which do not depress you, but instead slow down the nervous system.

Operations

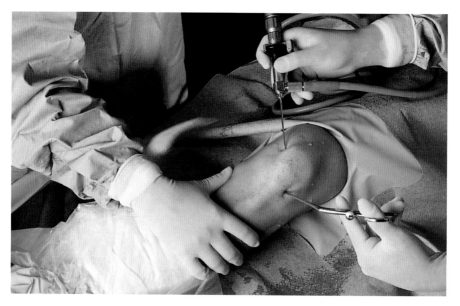

▲ *Many operations are now performed using instruments that allow surgeons to see inside the body without making large cuts.*

A surgical operation is when a doctor cuts or opens up a patient's body to repair or remove a diseased or injured body part.

An anaesthetic is a drug or gas that either sends a patient completely to sleep (a general anaesthetic), or numbs part of the body (a local anaesthetic).

Minor operations are usually done with just a local anaesthetic.

Major operations such as transplants are done under a general anaesthetic.

Major surgery is performed by a team of people in a specially equipped room called an operating theatre.

The surgical team is headed by the surgeon. There is also an anaesthetist to make sure the patient stays asleep, as well as surgical assistants and nurses.

The operating theatre must be kept very clean to prevent an infection entering the patient's body during the operation.

In microsurgery, a microscope is used to help the surgeon work on very small body parts such as nerves or blood vessels.

In laser surgery, the surgeon cuts with a laser beam instead of a scalpel, and the laser seals blood vessels as it cuts. It is used for delicate operations such as eye surgery.

An endoscope is a tube-like instrument with a camera at one end. It can be inserted into the patient's body during an operation to allow surgeons to look more closely at body parts.

▼ *In some operations, laser beams are used instead of a standard surgical knife. They allow more control and precision and reduce the risk of damage or bleeding.*

Transplants

▼ *A heart-lung machine is vital during a heart transplant as it keeps the blood circulating around the body and through the brain, keeping the body alive.*

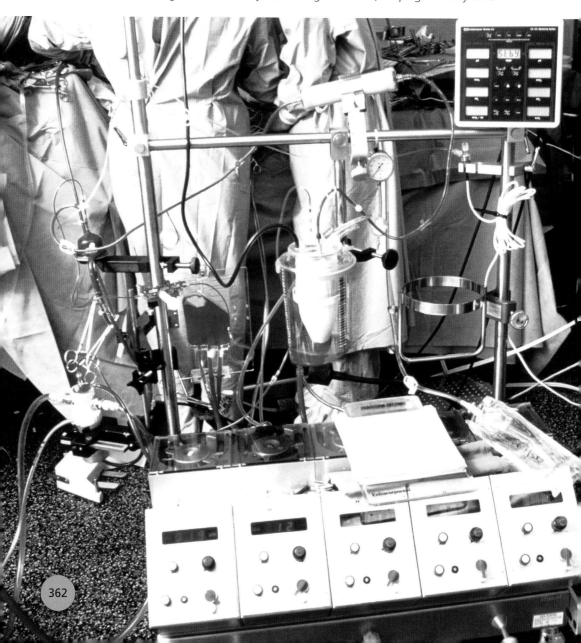

- **More and more body parts** can now be replaced, either by transplants (parts taken from other people or animals) or by implants (artificial parts).

- **Common transplants include**: the kidney, the cornea of the eye, the heart, the lung, the liver and the pancreas.

- **Some transplant organs** (such as the heart, lungs and liver) are taken from someone who has died.

- **Other transplants** (such as the kidney) may be taken from living donors.

- **After the transplant organ** is taken from the donor, it is washed in an oxygenated liquid and cooled to preserve it.

- **One problem with transplants** is that the body's immune system identifies the transplant as foreign and attacks it. This is called rejection.

- **To cut down** the chance of rejection, patients may be given drugs such as cyclosporin to suppress their immune system.

- **Heart transplant operations** last four hours.

- **During a heart transplant**, the patient is connected to a heart-lung machine that takes over the heart's normal functions.

Therapies

Therapies are used to help someone live a normal life. They can be used by themselves or with drugs or surgery.

Physiotherapy (or physical therapy) is usually used to make muscles and joints stronger.

Some physiotherapists work with people who have sport injuries or who have just had an operation, helping them get fit again.

Other physiotherapists work with elderly people or people who have been ill for a long time, helping them get well enough to live comfortably.

People who have difficulty breathing properly may have a specific form of physiotherapy to help clear their lungs.

Speech therapists help children and people who have difficulty speaking learn to speak and communicate.

◄ *Physiotherapy is used to help a patient recover strength in muscles and joints after an injury or illness.*

▲ *During speech therapy, a child may be encouraged to form sounds while seeing how his or her lips move in a mirror.*

- **Occupational therapists** help people cope with everyday activities.

- **Psychological therapies** are used to help treat people with mental health problems, such as depression.

- **Many psychological therapies** involve talking about your feelings to a therapist.

- **The therapist** helps you to understand why you feel the way you do and how to change the way you feel.

Complementary therapies

- **A complementary therapy** is a traditional treatment that is used as well as normal treatment to help to cure an illness or injury.

- **Many complementary therapies** have been used for hundreds of years.

- **Some complementary therapies** have been shown to work by scientific studies; others appear to work but there is no scientific evidence to prove it.

- **Two therapies** that are commonly used and have been shown to work are chiropractic and acupuncture.

▼ Traditional acupuncture has been practised in China for thousands of years and is often used to treat pain in muscles and joints.

Chiropractic is usually used to treat back and neck pain and many doctors will suggest chiropractic treatment for someone who has had back pain for a long time.

The chiropractor moves or manipulates the bones of the backbone and other joints.

Acupuncture is a traditional Chinese treatment.

During acupuncture very fine needles are pushed into the skin to ease pain and help cure illnesses.

Other complementary therapies that are commonly used but have not been proved to work include homeopathy, which uses tiny quantities of substances to treat illnesses and reflexology, which uses pressure on areas of the feet.

Complementary therapies do not replace normal treatment.

▶ Reflexology uses pressure points on the foot to treat illnesses throughout the body.

367

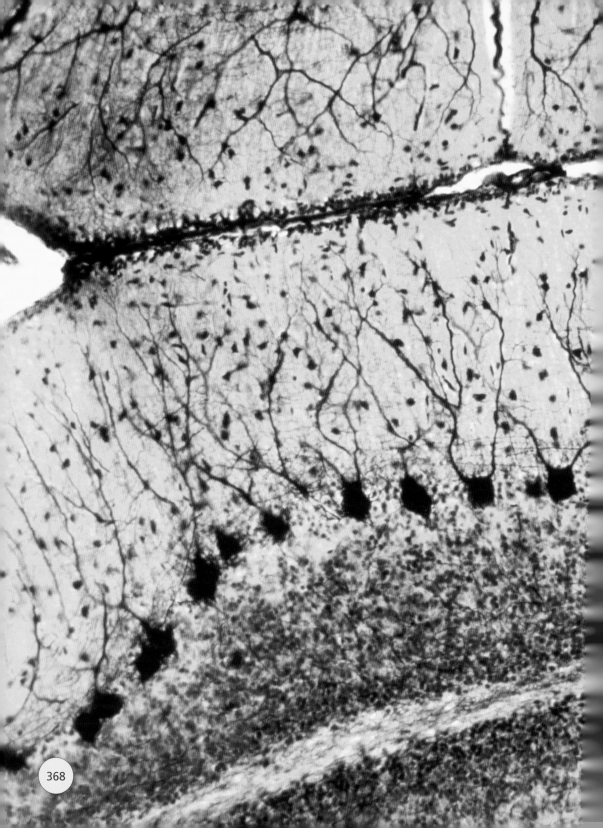

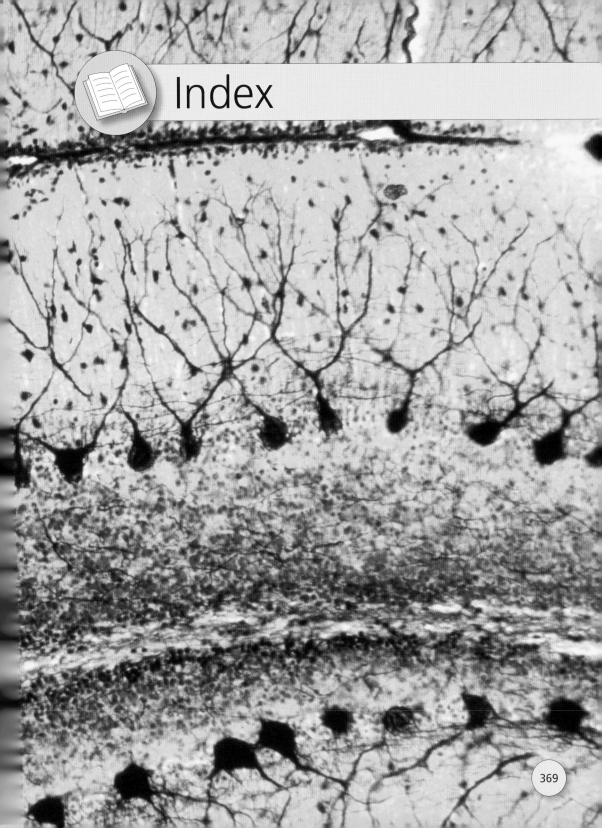

Index

Acknowledgements

All artwork is from the Miles Kelly Artwork Bank

The publishers would like to thank the following sources for the use of their photographs:

Cover (front) Springer Medizin/Science Photo Library, (spine) Shutterstock.com, (back) Ralf Juergen Kraft/Shutterstock.com

Dreamstime.com 24 Shippee; 38 Spanishalex; 41 Olga_sweet; 50 Danny666; 51 Scrappinstacy; 67 Rognar; 71 Mitchellgunn; 87 Razvanjp; 89 Choreograph; 101 Tupungato; 111 Fragles; 113 Banannaanna; 118 Aeolos; 127 Jabiru; 131 Rebeccapicard; 134 Kojoku; 136 Velkol; 147 K.walkow; 148 Diademimages; 149 Nruboc; 151 Achilles; 163 Ronibgood; 165 Khz; 172 Stratum; 202 Eraxion; 207 Tobkatrina; 209 Ligio; 218 Photoeuphoria; 219 Digitalpress; 253 Stratum; 264; 267 Barsik; 268 Jorgeantonio; 273 Jelen80; 274 Jelen80; 279 Doctorkan; 281 Alangh; 286 Kati1313; 288 Assignments; 289 Matka wariatka; 292 Keeweeboy; 312 Cb34inc; 318-319 Egal; 320 Kati1313; 325 Rachaelr; 328 Rognar; 329 Pro777; 331 Afagundes; 332 Tass; 333 Monkeybusinessimages; 352 Hanhanpeggy; 354 nyul; 367 Nataq

Fotolia.com (Bars)V. Yakobchuk; 12-13 Yuri Arcurs; 15 Sebastian Kaulitzki; 17; 28 Yuri Arcurs; 49 Olga Lyubkina; 69 chrisharvey; 85 Andy Dean; 110 Andreas Meyer; 120 Le Do; 139 Kiam Soon Jong; 138 Alexander Yakovlev; 245 Andrew Bruce; 300 Ronen; 311 Tein; 314 Celso Pupo; 318-319 Elena Schweitzer

Getty 175 Clouds Hill Imaging Ltd, Corbis Documentaryl; 359 Alfred Eisenstaedt, The LIFE Picture Collection

iStockphoto.com 19 Chris Dascher; 36 Seb Chandler; 47 technotr; 63 Raycat; 62 weicheltfilm; 93 annedde; 112 Maica; 130 deepspacedave; 132 ZoneCreative; 145 ZoneCreative; 146 arlindo71; 162 nyul; 168 Enjoylife2; 193 AvailableLight; 228 hartcreations; 229 Caziopeia; 254 robeo; 270 track5; 289 bonniej; 290 Bennewitz; 291 khilagan; 293 wbritten; 297 bloodstone; 298 shaunl; 323 Jason_V; 327 DNY59; 326 YinYang; 346 Fotosmurf03; 348 zlisjak; 349 peepo; 351 dgrilla; 364 Bryngelzon; 366 Yuri_Arcurs

Science Photo Library 25 Medical RF.com; 26 Zephyr; 33 Eye of Science; 35 Look at Sciences; 66 Paul Rapson; 99 Peter Gardiner; 109 Thomas Deerinck, NCMIR; 115 Anatomical Travelogue; 125 Kairos, Latin Stock; 135 David Mack; 140 Sovereign, ISM; 159 Zephyr; 178 Medical RF.com; 187 BSIP, Laurent; 188 Russell Kightley; 191 Gavin Kingcome; 194 Medical RF.com; 211 John Bavosi; 226 John Daugherty; 236 Jacopin; 239 Medical RF.com; 242 Professors P. Motta & F. Carpino/University 'La Sapienza', Rome; 250 Steve Gschmeissner; 259 Christian Darkin; 261 Christian Darkin; 262 Roger Harris; 277 Professors P.M. Motta & J. Van Blerkom; 278 Callot; 294 BSIP, JACOPIN; 303 Edward Kinsman; 337 Biomedical Imaging Unit, Southhampton General Hospital; 341 Hybrid Medical Animation; 342 Russell Kightley; 356-357 Zephyr; 362 Alexander Tsiaras; 365 Colin Cuthbert

Shutterstock 1 CLIPAREA | Custom media; 2–3 Sebastian Kaulitzki; 85 Deltaimages

All other photographs are from: Corel, digitalSTOCK, Image State, PhotoAlto, PhotoDisc

Every effort has been made to acknowledge the source and copyright holder of each picture. Miles Kelly Publishing apologizes for any unintentional errors or omissions.